THE PHYSICAL PROPERTIES
OF GLASS

THE WYKEHAM SCIENCE SERIES

General Editors:

PROFESSOR SIR NEVILL MOTT, F.R.S.
Emeritus Cavendish Professor of Physics
University of Cambridge

G. R. NOAKES
Formerly Senior Physics Master
Uppingham School

The aim of the Wykeham Science Series is to introduce the present state of the many fields of study within science to students approaching or starting their careers in University, Polytechnic, or College of Technology. Each book seeks to reinforce the link between school and higher education, and the main author, a distinguished worker or teacher in the field, is assisted by an experienced sixth form schoolmaster.

THE PHYSICAL PROPERTIES
OF GLASS

D. G. Holloway–University of Keele

 WYKEHAM PUBLICATIONS (LONDON) LTD
(A MEMBER OF THE TAYLOR & FRANCIS GROUP)
LONDON AND WINCHESTER
1973

First published 1973 by Wykeham Publications (London) Ltd.

Cover illustration—Tensile fracture surface of a glass rod.

ISBN 0 85109 320 5

Printed in Great Britain by Taylor & Francis Ltd.
10–14 Macklin Street, London, WC2B 5NF

Distribution:

UNITED KINGDOM, EUROPE, MIDDLE EAST AND AFRICA
Chapman & Hall Ltd. (a member of Associated Book Publishers Ltd.) 11 New Fetter Lane, London, EC4P 4EE, and North Way, Andover, Hampshire.

WESTERN HEMISPHERE
Springer-Verlag New York Inc., 175 Fifth Avenue, New York, New York 10010

AUSTRALIA AND NEW GUINEA
Hicks Smith & Sons Pty. Ltd., 301 Kent Street, Sydney, N.S.W. 2000.

NEW ZEALAND AND FIJI
Hicks Smith & Sons Ltd., 238 Wakefield Street, Wellington.

ALL OTHER TERRITORIES
Taylor & Francis Ltd., 10–14 Macklin Street, London, WC2B 5NF.

PREFACE

The properties of matter in bulk are determined ultimately by the nature and arrangement of the constituent atoms and the interactions between them. This book is an attempt to give a simple account of the physical properties of glasses and to show how, or sometimes how far, the properties of these materials can be understood using simple models to represent the behaviour of the constituent atoms. Inevitably this involves the introduction of fundamental concepts which are applicable to crystalline solids as well as to glasses and the opportunity has been taken to compare the observable, macroscopic properties of crystals and glasses. Thus, hopefully, the book may also help to provide a qualitative introduction to the physics of solids in general. The material treated here is complementary to that contained in two previously published books in this series, which deal with metals and with polymers.

Except for a few physical quantities such as those involved in technical definitions which have not yet been revised, SI units and the symbols recommended by the Royal Society (in *Quantities, Units and Symbols*, 1971) are used throughout this book as for the others in the series. A summary of the main symbols used appears on page 210. The scheme adopted for indicating the units on the axes of graphs and in tables also follows the recommendation of the Royal Society. The label on the axis of a graph or at the head of a column of figures is the quotient of the physical quantity and the unit used, so that all the figures can be regarded as pure numbers. However, we have chosen to indicate the physical quantities in words and the units in symbols. Thus, temperatures in degrees Celsius and in kelvins are distinguished by the labels Temperature/°C and Temperature/K, respectively.

The data used to construct some of the figures and tables have been collected from a wide variety of sources over many years and, even if it were possible, it would be impractical here to identify all these sources individually. I would, nevertheless, like formally to acknowledge these contributions from the work of many individuals and also indirectly from the glass industry at large. I am indebted to Messrs. E. Marriott, J. Sunderland and M. Cheney for help with some of the illustrations, and to Miss K. B. Davies for typing the manuscript.

v

My thanks are also due to Mr. P. N. Homer and Mr. C. R. W. Liley of the Advance Research and Technology Unit of J. A. Jobling & Co. Ltd., for the electron micrographs in figures 1.13 and 7.6.

Finally, I would like to express my appreciation of the advice, helpful criticism and suggestions offered by my collaborator, Mr. D. A. Tawney and, although I did not always need them, by my wife.

Keele   D. G. HOLLOWAY
October, 1972

CONTENTS

INTRODUCTION

GLASS is now one of the most common and also one of the most versatile materials made by Man ; many millions of tons are produced annually. It is not known where, when or how glass was first manufactured, although there is little doubt that by 1500 B.C. the ' industry ' was established. Archaeological discoveries have revealed that simple decorative glass objects and vessels were quite common in the Mediter-ranean civilizations of that period. Most of the important processes for shaping glass at high temperatures, including blowing, drawing, moulding and casting, were well established by the third century A.D. and the techniques for making and shaping articles of glass spread throughout the Roman Empire.

Glasses are also found in Nature. The mineral obsidian, usually a dark reddish-brown, translucent material, is a natural glass formed when molten rock cooled rapidly ; it has a chemical composition not very different from ordinary window glass. Foamed glass (pumice) and fibrous glass are also found near volcanoes ; some meteoritic particles are glassy and small glass pellets have been discovered in the samples of lunar ' soil ' brought back to Earth during the Apollo programme. Objects, such as knives, spear- and arrow-heads, fashioned from obsidian, have been found wherever this mineral occurs so that in one sense the art of manipulating glass goes back to the Stone Age.

The art of glass-making, like those of metallurgy and pottery, evolved over thousands of years. These arts all present particularly difficult problems in applied science ; only very recently has it become possible to explain why some of the recipes developed by trial and error are efficacious and even now the scientific understanding of glasses lags far behind that of metals. Although modern science and engineering technology have contributed enormously to the scale and efficiency of the glass industry, some of the most basic questions concerning the nature of the material it produces still cannot be answered with any confidence. The lack of basic understanding is perhaps reflected in arguments about the definition of a glass ! Glass technologists usually accept the definition proposed in 1945 by the American Society for Testing Materials : ' glass is an inorganic product of fusion which has cooled to a rigid condition without crystallizing '. Scientists interested in the structure and properties of materials are more likely to define a glass as a hard solid in which the arrangement of atoms or molecules is

ix

irregular in contrast to the highly-ordered arrangement in normal crystalline solids ; a glass is then a sort of ' petrified ', immobile liquid. This definition would encompass many organic materials, like glucose, Perspex and polystyrene for example ; the clear, hard, brittle form of toffee would be a glass while fudge is the crystallized form of essentially the same substance.

In this book we shall be concerned, for the most part, with the ' product of fusion ' which is made commercially, would be recognized by the layman as glass, and which consists of a solution of *inorganic oxides* one in another. Compositions are normally expressed in terms of the proportions, either by weight or by numbers of molecules, of the different oxides present, and for most of these glasses the major constituent is silica (SiO_2). Although the compositions of the glasses produced by different manufacturers for a particular application are much the same, in fact glasses are not compounds but *mixtures*. The proportions of the oxide constituents can be varied and this changes many of the physical and chemical properties.

Silica glasses offer a most attractive *combination* of properties ; they are transparent, hard, rigid solids which resist abrasion and scratching ; they are impervious to gases and liquids ; they are chemically inert and also will withstand high temperatures ; complex shapes can be fabricated quite simply and separate pieces can be joined to make a single homogeneous article. The raw materials required to produce them are among the most abundant in the Earth's crust and the ease of fabrication permits the use of continuous-flow, mass-production techniques. The major limitation is probably that glasses are brittle and in practice, although strong in compression, they fracture under very low tensile stresses. Perhaps in everyday experience we are made aware of this characteristic so often that the combination of other properties is taken for granted. The combination is, however, unique ; we might imagine an advanced society which does not have glass windows and bottles, but what else could be used for the envelope of a light bulb ? How could science, medicine and the technologies develop without telescopes, microscopes and spectrometers ?

The main purpose of this book is to show how the properties of glasses can be understood, at least qualitatively, in terms of their structure and the fundamental principles of physics. Discussion of the various physical properties will involve many of the basic ideas and some of the simpler models first developed and used to account for the corresponding properties of crystalline solids. The melting of glasses is described briefly in Chapter 1 but we shall not treat in any detail the technology of glass manufacture or the methods used in shaping articles from glass. However, figs. 1 and 2 may help to give a general impression of the variety of processes in use. There are several introductory books, and an excellent exhibition in the National

Museum of Science and Industry, South Kensington, which describe the techniques of production more fully.

Pressing

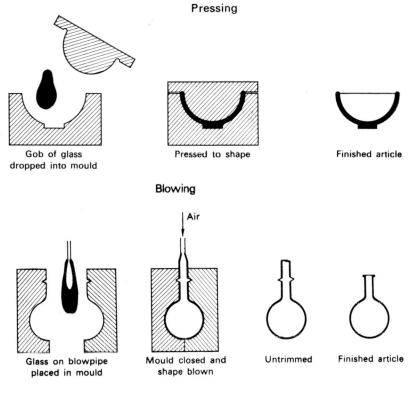

Gob of glass dropped into mould   Pressed to shape   Finished article

Blowing

Air

Glass on blowpipe placed in mould   Mould closed and shape blown   Untrimmed   Finished article

Pressing and blowing

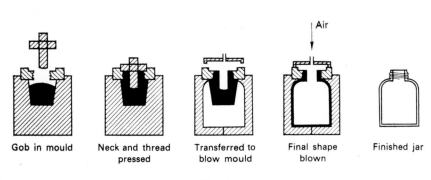

Air

Gob in mould   Neck and thread pressed   Transferred to blow mould   Final shape blown   Finished jar

Fig. 1.

Float process for plate glass

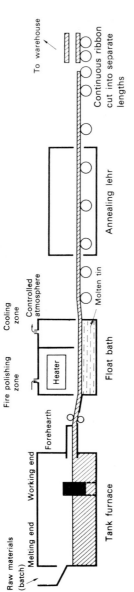

Some glass fibre processes

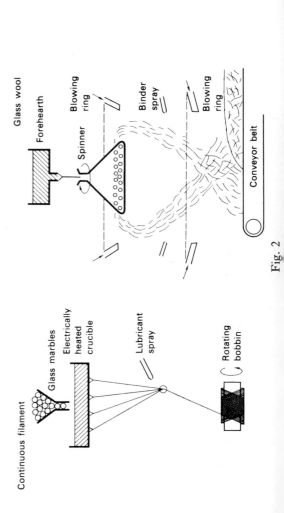

Fig. 2

CHAPTER 1

the composition and structure of glass

1.1. *Introduction*

GLASSES are fused mixtures of inorganic oxides; they do not have uniquely fixed compositions and many thousands of glasses, each with a different composition, are produced commercially. The vast bulk of these can, however, be considered in three main groups : the soda–lime–silica, the borosilicate and the lead silicate glasses. The soda–lime–silica glasses, which are mixtures of silica (SiO_2), sodium oxide (Na_2O) and calcium oxide (CaO), form by far the largest group of glasses in terms of tonnage produced and include the common sheet and plate glasses used for windows as well as the glasses used for most bottles and jars. Fused mixtures of mainly silica sand and boric oxide produce borosilicate glasses while silica sand, potash and lead oxide give the lead silicate glasses. Borosilicate glasses have a high chemical durability, high electrical resistivity and low thermal expansivity ; materials of this group are in common use for domestic ovenware and for laboratory glassware. Lead silicate glasses have a very high electrical resistivity and refractive index. Relatively costly raw materials and melting difficulties make this group of glasses expensive to produce, but nevertheless they are used for special purposes : for decorative, high-quality tableware, where the very high refractive index gives the classical ' sparkle ' when prismatic cuts are used to decorate the surfaces, and for the glass-to-metal seals in the bases of electric lamps and electronic valves.

All these glasses are produced on an industrial scale by the fusion of mixtures of readily available minerals or commercial heavy chemicals. For example, in the soda–lime–silica glasses calcium oxide is introduced as limestone ($CaCO_3$) and sodium oxide as sodium carbonate or nitrate (Na_2CO_3 or $NaNO_3$). The proportions of the raw materials in the batch are adjusted so that the required ratio of the oxides is formed during the fusion of the mixture. In addition all glasses contain small quantities of a wide variety of elements, some of which can play a major part in determining the physical or chemical properties of the glass. These minor constituents may have been present as impurities in the raw materials used to provide the major oxides or may have been deliberately introduced in order to influence the properties of the glass, as for example in the production of coloured glasses.

1

Composition (weight per cent)

Glass	SiO$_2$	Na$_2$O	K$_2$O	CaO	MgO	B$_2$O$_3$	Al$_2$O$_3$	Fe$_2$O$_3$	PbO
1. Soda–lime–silica (plate glass)	72·5	13·0	0·3	9·3	3·0		1·5	0·1	
2. Soda–lime–silica (bottles, jars)	73·0	15·0		10·0			1·0	0·05	
3. Soda–lime–silica (electric lamp bulbs)	73·0	16·0	0·6	5·2	3·6		1·0		
4. Borosilicate (Pyrex)	80·6	4·2		0·1	0·05	12·6	2·2	0·05	
5. Lead–silicate (tableware)	55·5		11·0						33·0
6. Lead–silicate (electrical : seal to lead-in wires, lamp bulbs and valve bases)	63·0	7·6	6·0	0·3	0·2	0·2	0·6		21·0
7. Lead–silicate (radiation shields)	30·0		3·0						65·0
8. Aluminosilicate (electrical, high durability glass fibres)	54·6	0·6		17·4	4·5	8·0	14·8		
9. Aluminosilicate (combustion tubes, ' top-of-stove ' ware)	55·3	0·6	0·4	4·7	8·5	7·5	23		

Table 1.1. Typical compositions of commercial glasses.

2

Table 1.1 shows the compositions of some glasses produced commercially for various purposes. Most of these glasses will contain trace quantities of other elements such as arsenic (As) and antimony (Sb) as well as SO_2, H_2O and other dissolved gases. While in principle the variety of possible glasses is unlimited, in practice glasses produced by different manufacturers are often very closely similar in composition. The requirements of (i) easy melting on an industrial scale, (ii) a wide range of temperature within which the glass can be shaped by a given process, and (iii) good durability, plus limitations in the nature of available economical raw materials, often lead to very much the same basic mix for producing a glass of given type for a particular application.

1.2. Glass melting

Glasses are melted on an industrial scale either in ceramic pots holding up to about 15 cwt of molten glass, or in large tanks constructed from special refractory materials. In pot-melting, more than a dozen pots may be placed in a large furnace at the same time or in sequence and once melting is complete individual pots can be removed to enable the glass to be worked from the pot. The larger tanks are an integral part of the furnace and many large ones holding over 1000 tons of glass are operated continuously, unmelted materials being fed in at one end while glass is withdrawn continuously from the other.

Glass is rarely melted from a batch consisting entirely of fresh raw materials ; the charge for the pots or the material fed into the melting-tank consists of a mixture of new raw materials in the correct proportions and scrap glass, known as cullet. In nearly all glass-forming processes (i.e. the rolling, drawing, pressing, blowing or other procedures that are used to shape the hot viscous glass into a product) there is some rejected or waste material, often because some of the material has to be removed from the shape originally formed in order to produce the finished article. The use of cullet in the batch is clearly economical but it also serves an important role in aiding the reactions and solution processes that occur during melting. The proportion of cullet to fresh raw material is typically 15–30 per cent.

During the glass-melting process when the batch is raised to 1350°–1600°C, depending on composition, many complex physical and chemical changes occur although the temperatures are well below the individual melting points of some of the batch constituents. The initial changes include the evaporation of water contained in the batch as water of crystallization and the decomposition of carbonates, sulphates, etc. with the release of CO_2, SO_2, SO_3, etc. Some of the raw materials melt as the temperature rises and the liquids formed begin to dissolve the other more refractory constituents. The final stage of melting, known as *refining* the glass, involves the removal of residual bubbles of gas (seed) liberated by the raw materials and trapped in the highly

3

viscous molten solution of oxides. This stage generally takes much longer than the conversion of the original mixture of solids into a substantially homogeneous liquid phase. One of the older methods of accelerating this process was to plunge moist materials (potatoes or green wood !) into the glass when the violent release of large amounts of water vapour helped to sweep the smaller bubbles to the surface. The effect is now more usually achieved by the addition of chemical refining agents which will decompose, liberating large bubbles, during the later stages of melting.

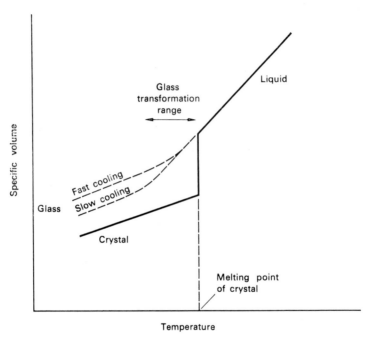

Fig. 1.1.   Transformations from liquid to crystalline, and liquid to glassy states

The complete melting process may take up to 24 hours or even longer and during this prolonged time at high temperature small but sometimes important changes can occur in the composition of the glass due to volatilization of some of the constituents and to reactions with the walls of the container and with furnace gases. In large continuous tank-furnaces the temperatures in each section are adjusted so that mutual solution, mixing by convection, and refining occur as the glass progresses along the length of the tank ; the final section brings the glass to the temperature at which it has the correct viscosity for working.

4

1.3. *Nature of the glassy state*

At the usual melting temperatures glasses are highly viscous liquids ; typically the viscosity is $\sim 10^2$ poise (~ 10 N s m^{-2}), about the same as glycerol at $0°C$. As they cool their viscosity increases progressively and continuously (see fig. 1.12). At $\geqslant 10^8$ poise measurements are made by twisting a rod of material and above 10^{15} poise it becomes impossible to obtain meaningful values, for the material is as rigid as normal crystalline solids. This continuous transition from a liquid to a solid condition provides a distinction between glasses and crystalline solids. The continuous nature of the transition is apparent in other physical properties which are related to the internal structure and a discussion of the implications will lead us on to a model for the structure of glasses.

In fig. 1.1 we have illustrated the changes in specific volume (volume per unit mass) that would occur as a simple, imaginary substance is cooled from the liquid state so as to form (*a*) a crystalline solid and (*b*) a glass. When the liquid is cooled very slowly an abrupt change occurs in the specific volume and in other physical properties at a characteristic temperature, the melting point of the substance, where the material crystallizes. When the liquid is cooled rapidly there is no abrupt change in specific volume at any temperature, instead the slope of the specific volume temperature curve changes continuously over a range of temperature. At low temperatures the rate of change of volume with temperature (i.e. the thermal expansivity $\alpha = V^{-1}\,\mathrm{d}V/\mathrm{d}t$) is similar for both the glass and the crystalline material, but the absolute magnitude of the specific volume of the glass is much larger and it also varies with the rate at which the original liquid was cooled. The transition from the liquid to the glassy state takes place over a range of temperature ; there is no clearly defined transition temperature which could be compared with the melting point of a crystalline solid.

If the original liquid were more complex, such as a mutual solution of two metals rather than the simple substance depicted in fig. 1.1, then the crystallization process would be more complex. Instead of a single melting temperature we usually find a range of temperatures over which crystallization occurs progressively and the composition of the crystallizing material changes as the temperature falls through the range. Nevertheless we can identify two specific temperatures (at a given pressure) which would be characteristic of the composition of the mixture : the *liquidus temperature*, above which no crystals exist and the *solidus temperature*, below which no liquid exists.

There is a further and perhaps still sharper distinction between the continuous transition from liquid to glassy states and the crystallization which may occur when a simple liquid is cooled below its melting point : there is no evolution of ' latent heat ' during the liquid to glass transformation. The latent heat of a transformation is associated with the

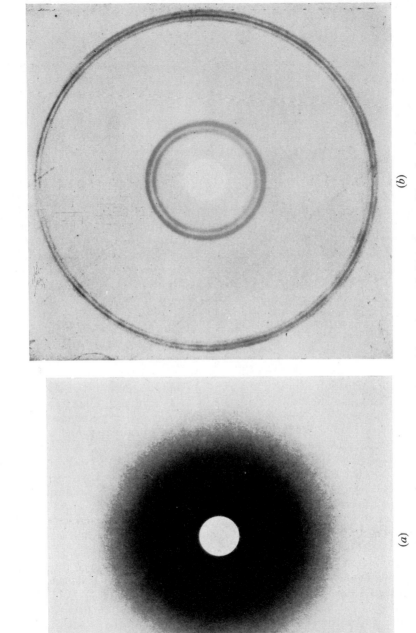

(a)

(b)

Fig. 1.2. X-ray diffraction patterns from (a) a glass and (b) polycrystalline aluminium.

abrupt change in the disorder, which occurs when a liquid crystallizes and the atoms or molecules change from an irregular arrangement in space to a highly-ordered, regular one. Both the continuous nature of the transformation and the absence of any latent heat suggest that the molecular arrangement in glasses is akin to that in liquids. This is indeed borne out by direct experimental investigation of the structure using the diffraction of X-rays or neutrons. A glass yields a very broad diffuse diffraction ring comparable with the patterns produced by liquids and quite unlike the sharp rings obtained from a poly-crystalline solid. X-ray diffraction patterns from a soda–lime–silica glass and aluminium are illustrated in fig. 1.2. The atoms or molecules in a crystal lie on regularly spaced sets of planes. When the crystal is held in an X-ray beam, each of these planes in effect acts as a partial reflector, although a strong beam is reflected from the whole crystal only if the individual reflections from successive planes reinforce one another ; that is if the path difference between the rays ' reflected ' by successive planes is equal to a whole number of wavelengths. It is this condition which leads to the well-known Bragg Law

$$2\,d\sin\theta = n\lambda, \quad n = 1, 2, 3, \text{ etc.}$$

which relates the angle, θ, at which a strong diffracted beam occurs, the interplane spacing, d, and the wavelength of the X-rays, fig. 1.3.

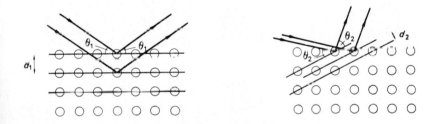

Fig. 1.3.   Bragg reflection of X-rays from crystal planes.

The absence of sharp, strongly diffracted beams in the X-ray or neutron diffraction patterns from glass indicates that there are no well-defined planes in the structure on or around which the constituent atoms are regularly arranged. In a general way we imagine a glass to be a random arrangement of molecules much like a liquid with the very important difference that, below the transformation region, the molecules are much less mobile. Although the individual atoms vibrate, due to thermal agitation, about their mean positions, in solid glass only very rarely will they be able to exchange positions. The spacing between adjacent molecules will be much the same in the glassy and the crystalline form of the material but the regular repetition of the spatial arrangement of groups of molecules will be absent in the glass. The structure of a glass may be like a random jumbled pile of balls rather than the familiar regular stacks, such as in fig. 1.4, used to represent the arrangement of atoms in simple crystals.

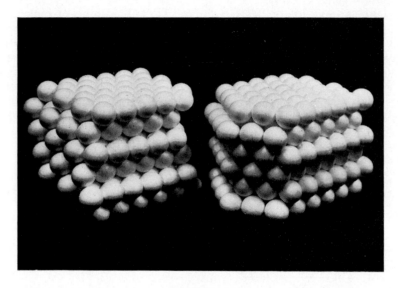

Fig. 1.4. Ball models used to illustrate the atomic arrangement in single crystals.

Only those liquids which have very high viscosities at their melting point can form glasses when cooled at normal rates. The viscosity is related to the ease with which the molecules in the liquid can move relative to one another. If the adjustments in position required to bring molecules into a configuration corresponding to that of the regular crystalline arrangement are to be inhibited when the liquid is cooled below the melting point, molecular rearrangement must be a slow process and thus the viscosity will be high. Below the melting point the crystalline form of the material invariably has the lower energy (strictly a lower *free energy* in thermodynamic terms) and is therefore the true equilibrium form of the material. If a glass is held at a temperature in the transformation range and below the liquidus temperature for a very long time, crystals will begin to grow. The glass is said to *devitrify*. It is largely the difficulty of rearranging the molecules which precludes the attainment at normal cooling rates of the true, minimum-energy, crystalline configuration. As the temperature falls below the transformation region, molecular rearrangement becomes increasingly difficult and for useful glasses significant rearrangement is impossible at normal temperatures : the glass is stable although not in thermo-dynamic equilibrium.

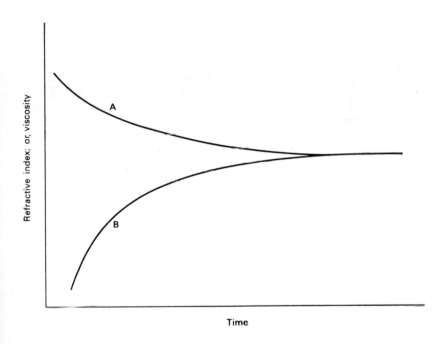

Fig. 1.5. Approach to a metastable equilibrium in the transformation range of temperature. The change in refractive index and viscosity with time at a given temperature T is illustrated for two samples of the same glass ; (A) previously stabilized at a temperature $< T$ and (B) at a temperature $> T$.

For many materials that can be cooled to a glassy state it appears that the molecular configurations can achieve a metastable or ' pseudo-equilibrium ' form at any given temperature within the transformation region. This is indicated by the fact that the physical characteristics of the material such as the refractive index, the viscosity, the density,

9

etc. tend with time towards unique values, at any given temperature, which are the same whether the material has previously been held at a higher or a lower temperature, fig. 1.5. The lower the selected temperature the longer it takes to reach a steady ' pseudo-equilibrium ' condition because the rate of molecular rearrangement falls very rapidly with the temperature, until finally no perceptible changes occur within a reasonable time and the configuration is effectively ' frozen in '. The temperature in the transformation region at which the structure of a given piece of glass would be stable is called the *fictive temperature*. For a particular piece of glass this is the temperature within the transformation range at which the prevailing atomic configuration is in pseudo-equilibrium and over shorter periods of time no changes in configuration would occur although some atomic movements are possible. The higher the fictive temperature the greater the difference between the physical properties of the sample of glass and the crystalline form of the material. In our simple example in fig. 1.1 the rapidly cooled sample has a higher specific volume, reflecting the more open structure corresponding to a higher fictive temperature ; the more rapid cooling results in a higher fictive temperature because this is the point at which the changes in molecular configuration fail to keep pace with the changing temperature.

1.4. *The structure of silica glass*

Over a long range, say $\gtrsim 1000$ atomic diameters, the atomic arrangement in a solid may be regular as in a crystal or irregular as in a glass. However, the *local* arrangement of atoms is usually regular and is determined by the nature of the atoms involved either through the character of the bonds formed or by the relative sizes of the atoms present. Where bonding is directional, as for example in the covalent (electron-sharing) bond, the local arrangement is fixed by the bond angles ; where bonding is ionic and therefore non-directional the local arrangement is fixed by the relative sizes of the cations and anions. Transfer of electrons from the electro-positive to the electro-negative atoms usually makes the resulting anions much larger than the cations, but ions of a given kind tend to keep a constant size so that, for example, the O^{2-} ion has the same radius whether it is in MgO, SiO_2 or TiO_2. We can picture the local packing of two dissimilar ions as being determined simply by the number of anions that can be placed in contact with the cation without overlapping one another ; the centres of the anions must be at least 2 radii apart, see fig. 1.6. In the complete structure we must preserve electrical neutrality : the ratio of the total numbers of positive and negative ions in the solid must be the same as the ratio of the ionic charges. However, this requirement influences the ways in which we can pack the local groups together rather than the structure of the local group itself.

10

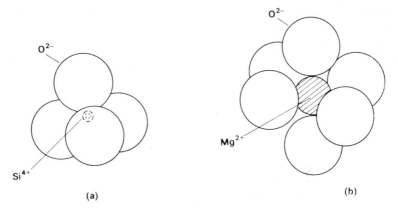

Fig. 1.6. Local packing of ions in (*a*) silica and (*b*) magnesium oxide.

In crystalline forms of silica and in many silicates the silicon ion is associated with four neighbouring oxygen ions to form a tetrahedral structural unit ; the silicon ion lies at the centre of the tetrahedron and the oxygen ions at the four corners, fig. 1.7 *a*. In silica adjacent tetrahedra share a corner oxygen. The distance between the centres of the Si and O ions is 0·16 nm, the oxygen ion being very much larger ($\sim \times 3$) than the silicon ion. Strictly the Si–O bond is not a pure ionic bond (few are !) but involves some electron sharing. The tetrahedral distribution of the four oxygens around a silicon atom satisfies the requirements of both the directed bonding characteristic of covalency and the size ratio demanded by ionic bonding. It is customary to describe the structure and to discuss the properties of silicates and the silica based glasses as if the atoms were held together by ionic bonds.

The three different forms of crystalline silica, cristobalite, tridymite and quartz, correspond to different, regularly repeating, three-dimensional patterns of these tetrahedra. In the most widely accepted model for the structure of silica glass, the so-called *random network theory*, it is envisaged that these tetrahedra are randomly arranged in space. This structure is admirably suited to representation by 'ball and spoke' models. Suppose while building such a model we allow a flexibility in the relative orientation of adjacent tetrahedra, i.e. when joining two tetrahedra we allow some variation in the angle, θ, defined by the two lines, SO and S'O in fig. 1.7 *b*, joining the silicon ions to the common corner oxygen, and we also permit a relative rotation of the tetrahedra so that the right-hand tetrahedron in fig. 1.7 *b* can be rotated about the line S'O as axis, then a variation of a few per cent only in the Si–O–Si angle, θ, will produce a three-dimensional structure which completely lacks the long-range order characteristic of a crystal

11

and yet preserves the nearest-neighbour separations and the orientations required by the bonds between the silicon and oxygen. Such model building is still being undertaken seriously in several laboratories in order to examine some of the details of the network structure. This

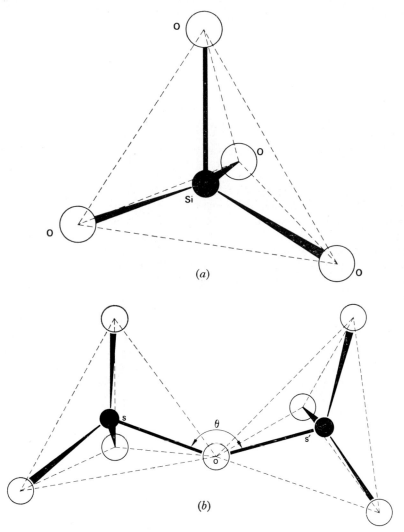

Fig. 1.7. (a) SiO₄ tetrahedral unit; (b) adjacent units sharing a corner oxygen ion. (Following the usual convention these diagrams show just the positions of the centres of the ions so that the arrangement can be seen clearly. Figure 1.6 a shows more faithfully the relative sizes of the ions.)

apparently naive approach contrasts with the sophistication of much of present-day science but is likely to be particularly useful in this kind of situation where it is very difficult to represent the conceptually simple model in conventional mathematical terms.

Illustration of the three-dimensional random network of tetrahedra in two dimensions is more difficult than actually building 'ball and spoke' models and it has become traditional to use a two-dimensional analogue in the same way as W. H. Zachariason did when he first proposed the random network model for the structure of glassy silica. Figure 1.8 illustrates in this way the structure of a crystal and of a glass according to the random network theory. In order to confine the structure to two dimensions we take as the structural unit a triangle with an atom, R, at the centre and three atoms, X, at the corners ; each corner atom is shared by an adjacent structural unit so that, with an array consisting of a large number of triangular units, each atom of R is effectively associated with one-half the number of corner atoms. Thus our imaginary two-dimensional material has the formula R_2X_3.

It was suggested, in the previous section, that the molecular arrangement in a glass was a ' frozen in ', static form of the random arrangement in the liquid. In keeping with this notion, fig. 1.8 b could also be taken to represent the arrangement of the RX_3 structural units in liquid R_2X_3 provided we regard it as an ' instantaneous picture ' and allow that the particular form will change with time. We may also note that the number of large ' holes ' in the random network formed by rings of RX_3 triangles will increase with temperature. It is this sort of change in the structure which is responsible for the large thermal expansion of a liquid. The mobility of the individual structural units disappears abruptly as the liquid crystallizes or decreases progressively throughout the transformation range as the material forms a glass. Further reductions in temperature lead thereafter to much smaller changes in volume, due almost entirely to the decrease in the average size of the individual triangles, or in three dimensions the tetrahedra, as the amplitude of vibration of the constituent atoms falls.

1.5. *The structure of soda–silica glass*

In a piece of pure silica glass each of the SiO_4 tetrahedra would share its four corner oxygen ions with neighbours. All the ions in the sample would be strongly bonded and would form part of a continuous random network : the piece of glass would be a single large molecule. Addition of monovalent ions in the form of an oxide such as Na_2O disrupts the continuity of this network. In a soda–silica glass the number of oxygen ions is more than twice the number of silicon ions so that it is not possible for every oxygen ion to be bonded to two silicons. Each silicon ion is tetrahedrally surrounded by four oxygens but not every oxygen ion can form part of two adjacent tetrahedra. The oxygen ions

13

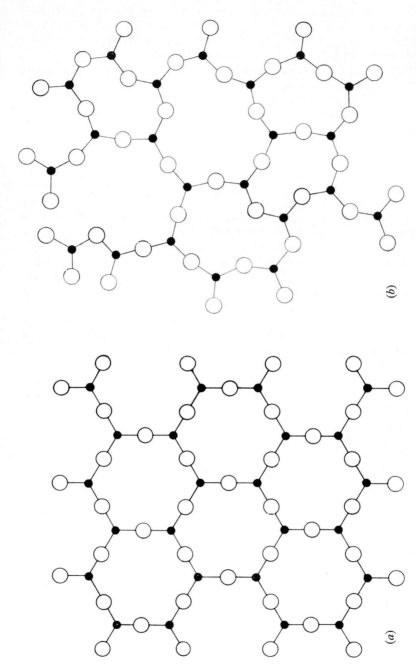

Fig. 1.8. Two-dimensional analogue of the structure of (a) the crystalline and (b) the glassy form of a substance, R_2X_3. The open circles represent atoms of X and the filled circles atoms of R.

14

which are bonded to only one silicon ion are said to be *non-bridging*. The sodium ions occupy some of the spaces in the network, near non-bridging oxygens so as to preserve local electrical neutrality, and are surrounded on the average by six oxygen ions. For each pair of mono-valent metal ions (and therefore one extra oxygen ion) introduced, one of the links in the network is broken and two non-bridging oxygen ions are produced.

The two-dimensional analogue of a soda–silica glass is illustrated in fig. 1.9 where we have added a proportion of M_2X to the basic network formed by the R_2X_3.

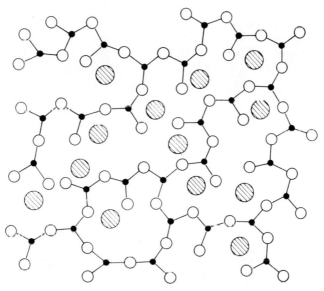

Fig. 1.9. Two-dimensional analogue of a modified glass ; R_2X_3 modified by the addition of some M_2X ; notation as in fig. 1.8, with shaded circles representing atoms of M.

The addition of Na_2O to silica breaks up the continuity of the network structure and hence produces quite marked changes in physical properties. At a given high temperature the viscosity of a soda–silica glass is much less than that of pure silica. Soda–silica glasses are therefore much ' easier ' to produce and to work, in the sense that the temperatures required to remove bubbles from the melt or to shape the glass are much lower. It is not normally possible to make a soda–silica glass with the molecular proportions of soda to silica greater than about 1 : 1. Above this ratio, *on average* more than two of the four oxygen ions surrounding each silicon would be non-bridging and the formation

15

of a continuous random network of SiO_4 tetrahedra would be impossible. The molecular alignments necessary to produce crystallization can then occur relatively easily and crystals will grow in the liquid as the temperature falls below the liquidus.

1.6. *Composition and structure of practical glasses*

Practical glasses contain a wide variety of oxides and the ideas of the random network theory have been used to characterize the roles which these play in the structure.

According to Zachariason the one essential ingredient in an oxide glass is a network forming oxide. SiO_2 is by far the most commonly used *network former*, but GeO_2, As_2O_5, P_2O_5 are all known to form tetrahedral structural units which share corners in crystalline compounds and all form glasses. B_2O_3 forms triangles and again can readily produce a glass.

The other classes of oxide used in glass-making are designated *network modifiers* if, like Na_2O, they disrupt the continuity of the network, or *intermediates* if they can either join the network or can occupy holes in the network. The alkali metal oxides and alkaline earth oxides

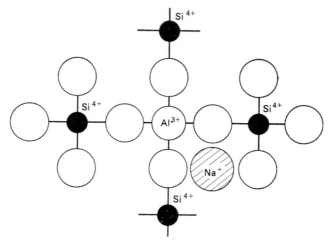

Fig. 1.10.   Schematic representation of Al_2O_3 in a silica network.

are network modifiers. Lithium and potassium oxide behave much like sodium oxide ; however, the lithium ion is smaller and the potassium larger than the sodium ion so that Li^+ will tend to occupy smaller holes in the network and K^+ larger ones than Na^+. Calcium and magnesium oxide also break up the continuous network but the divalent cations Ca^{2+} and Mg^{2+} each produce two non-bridging oxygen ions.

16

Oxides such as Al_2O_3, BeO, TiO_2, ZrO_2 are usually classed as intermediates because under certain conditions it appears that they can join the continuous network, although unable to form networks by themselves, but that they may also occupy the holes in between the SiO_4 tetrahedra.   In the case of Al_2O_3, for instance, the Al^{3+} ion will substitute for a Si^{4+} ion, i.e. an AlO_4 tetrahedron can be formed which will join in the network formed by the SiO_4 tetrahedra, provided an additional cation is available and can be located in a nearby hole to ' preserve the electrical neutrality, fig. 1.10.   The additional cation may be an alkali metal ion or even an alkaline earth ion " shared " between two adjacent AlO_4 tetrahedra.   Lead oxide is unusual, in that simple binary mixtures consisting of PbO and SiO_2 containing very large amounts of lead oxide, ~ 80 mole %†, readily form glasses.   It is thought that some of the Pb^{2+} ions form links between two SiO_4 tetrahedra by bonding to the corner oxygens and may therefore be considered as taking part in the formation of a network, albeit a much more open one than that formed by SiO_4 alone, fig. 1.11.

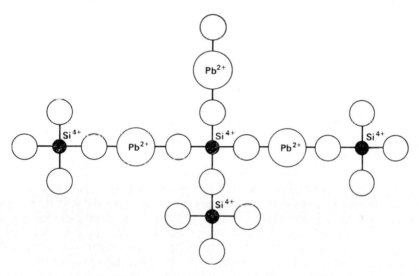

Fig. 1.11.   Schematic representation of PbO in a silica network.

The inorganic oxides commonly used in making glasses are listed in table 1.2, together with summaries of the more important effects which they have on the properties of the glasses.   We shall discuss the physical basis of many of these effects in later chapters.

† Mole % gives the percentile proportion of molecules in the mixture, wt % gives the percentile proportion by mass.

SiO_2	Basic glass-former (network former). Crystalline silica has very high melting point and liquid silica very high viscosity. High concentration of silica in a glass confers high softening temperature, low thermal expansion, good chemical durability.
B_2O_3	Network former. Will join network structure of silica glasses reducing viscosity without producing adverse changes in thermal expansion and durability. Is a component of all heat-resisting and ' chemical ' glasses.
PbO	*Not* a network former but can link SiO_4 tetrahedra. Widely used for high dielectric constant, refractive index, and resistivity. Expensive. Needs oxidizing furnace atmosphere.
Na_2O	Network modifier. Markedly lowers softening point. Raises thermal expansion and ionic conductivity. Reduces durability.
K_2O	Network modifier—similar to Na_2O but larger K^+ ion less mobile.
LiO_2	Network modifier—similar to Na_2O but smaller Li^+ ion more mobile. Promotes devitrification.
CaO	Network modifier. Inhibits mobility of alkali ions, hence increases resistivity and durability of alkali glasses. Shortens the working range.
$\left.\begin{array}{l} MgO \\ ZnO \end{array}\right\}$	Network modifier—as CaO.
BaO	Cheaper substitute for PbO.
Al_2O_3	Intermediate. Can join network in AlO_4 tetrahedra which are a different size from SiO_4. Strongly suppresses devitrification ; increases viscosity and therefore makes melting and refining more difficult.

Table 1.2.   Common constituents of glasses and their influence on properties.

1.7. *Characteristic reference temperatures*

As we have seen, for a glass there is no sharp discontinuous transition into the liquid state but a progressive decrease in viscosity as the temperature increases through the transformation range. Both the width and the position of this temperature range vary from one glass to another and selected temperatures within the range are used to characterize the glass. At one time the temperatures were defined in terms of convenient but arbitrary laboratory tests : for example, the *fibre softening point* was the temperature at which a fibre of specified size would elongate at a certain rate under its own weight. A series of reference conditions have recently been redefined in terms of the viscosity of the glass but so as to correspond roughly with the previous conditions.

The viscosity of glass in melting-tanks is of the order of 10^2 poise. When being pressed into moulds or drawn into tubing or rod the viscosity (η) is between 10^3 and 10^6 poise and arbitrarily the temperature at which $\eta = 10^4$ poise is called the *working point*. The *softening point* is the temperature at which $\eta = 10^{7 \cdot 6}$, at the *strain point* $\eta = 14 \cdot 5$ and at the *annealing point* $\eta = 13 \cdot 4$ poise. The strain point is the highest temperature from which the glass can be rapidly cooled without intro-

Reference	Viscosity/ poise	Temperatures/(°C)						
		Silica	96 per cent silica	Aluminosilicate (9)† (combustion tubes)	Pyrex (4)†	Soda–lime– silica (3)	High lead glass (7)	Lead glass (6)
Strain point	$10^{14\cdot5}$	~1000	820	670	520	470	300	390
Annealing point	$10^{13\cdot4}$	1100	900	720	565	510	320	430
Softening point	$10^{7\cdot6}$	1600	1500	940	820	700	380	620
Working point	10^4	—	—	1220	1220	1000	500	970

† These numbers correspond to the listing in table 1.1.

Table 1.3. Characteristic reference temperatures for various glasses.

ducing serious internal stresses and the annealing point is the temperature at which any internal stresses will be relieved in a few minutes. The non-integral indices in these new definitions for the reference temperatures arise because of the efforts to preserve roughly the same convenient reference points but nevertheless to express the conditions

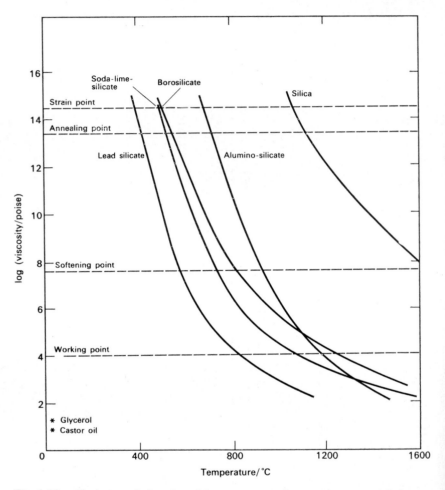

Fig. 1.12.   Variation of viscosity with temperature for some commercial glasses.

in more formal and quantitative terms.   The reference temperatures, definitions and some values for selected glasses are shown in table 1.3. The change in viscosity with temperature for some different glass types is illustrated in fig. 1.12.

20

1.8. *The production of silica glass*

Since commercial glasses are based on silica as the network former, it would appear logical and certainly in the traditions of physics to study in detail the structure and properties of this ' simplest ' system in the hope of establishing a basic theoretical framework for the explanation of the behaviour of the more complex glasses. To some extent this has been done, as we have already seen in the discussion of the network structure. However, it is still not possible to produce a very pure ' good ' silica glass in any quantity and in general it is therefore not feasible to base studies of the physical properties of silicate glasses in any detailed way on silica itself.

The production of silica glass involves very special techniques which have nothing in common with normal glass-melting practice. Not only are much higher temperatures required, mineral quartz crystals melt at temperatures $\gtrsim 1700°C$, but the liquid silica has a very high viscosity, $\sim 10^7$ poise (10^6 N s m^{-2}), comparable with plasticene ! The viscosity cannot be significantly reduced by increasing the temperature because volatilization becomes appreciable.

Silica glass has been produced commercially only since 1900 ; indeed, prior to the discovery of the oxy-hydrogen flame (1821), quartz was generally believed to be infusible. Silica glass is now produced by several different processes and the properties of the material vary with the method of production as well as with the source of raw material. Quartz sand is fused to make a silica glass using a large carbon rod as a heating element ; the rod is embedded in a large box filled with sand and an electric current of 1000–2000 A is passed through the rod to fuse a core of sand in the centre of the box. Silica glass prepared in this way has a large bubble content which makes it translucent rather than transparent. Fusion under vacuum can be used to produce a glass which is relatively bubble free and transparent. Alternatively, a variety of flame fusion or arc-melting procedures can be used in which purer crushed mineral quartz crystals are fed into the flame or the arc and fused silica droplets impinge on an adjacent surface to build up progressively a large mass of bubble-free material. A new type of silica glass exceptionally free from mineral impurities is now produced, particularly for optical components, from volatile silicon compounds such as $SiCl_4$. These volatile compounds can be purified easily by fractional distillation and then oxidized to form SiO_2 powder and fused electrically. Alternatively the volatile compound can be fed directly into a flame where it is oxidized and simultaneously fused to form droplets which fall onto the surface of previously fused material.

However, none of these products is an ideal, pure, glassy silica : those produced from quartz minerals retain most of the inorganic impurities present in the original material ; those produced from volatile ilicon compounds have very few metallic impurities but the concentration of chlorine or of OH ions is relatively high. In addition to these problems of purity there are significant structural differences, apart from bubble content, depending on the method of fusion : electrical fusion results in a structure which contains residues of the original crystalline order ; flames give more complete fusion and therefore result in a more random structure but introduce hydroxyl ions, which form Si–OH linkages and break up the continuity of the network. Samples of ' fused silica ' produced by different methods, or from different raw materials, have significantly different properties, although these differences are small compared with those between silica and the common glass types.

From a practical point of view fused silica is a very difficult glass to work with ; very high temperatures are required to shape the glass and even then the high viscosity restricts the kind of shapes that can be produced conveniently. A glass containing 96 per cent silica (and about 3 per cent B_2O_3 plus residual alkali oxides) which has properties similar to ' pure ' silica provides a very useful compromise and articles of this glass are produced by a very ingenious method which avoids the major difficulties involved in working with pure silica. The original procedure was developed and patented by the Corning Glass Works (U.S.A.). A glass containing a fairly high proportion of boric oxide and a few per cent of alkali oxide is first melted ; this has a relatively low softening point and is easily worked. Articles are made from this glass and are then heat-treated for some hours at 500–600°C. The composition of the original glass is specially chosen so that under prolonged heat treatment two distinct glass phases are formed (fig. 1.13), one consisting of almost pure silica and the other rich in B_2O_3 and alkali. After heat treatment the majority of the modifier oxide can be dissolved out by soaking in acid and this leaves a porous silica glass behind. This porous structure is consolidated by a final high temperature treatment (1000–1200°C) during which the article shrinks about 30 per cent by volume and the glass becomes impermeable.

1.9. *Criticism of the random network theory*

Our discussion of the structure of oxide glasses has been based on the random network theory of Zachariason. This theory has been of great value in understanding the nature of glass ; it has provided a conceptual model for the structure and hence a context for the discussion and explanation of the properties of glasses. We have already used it to ' explain ' the limit to glass-forming compositions in the simple case of soda–silica glasses and the same kind of argument can be used to understand the limitations which exist for more complex mixtures of oxides. The evidence for the random network theory is, however, not conclusive and in recent years many authors have argued that on

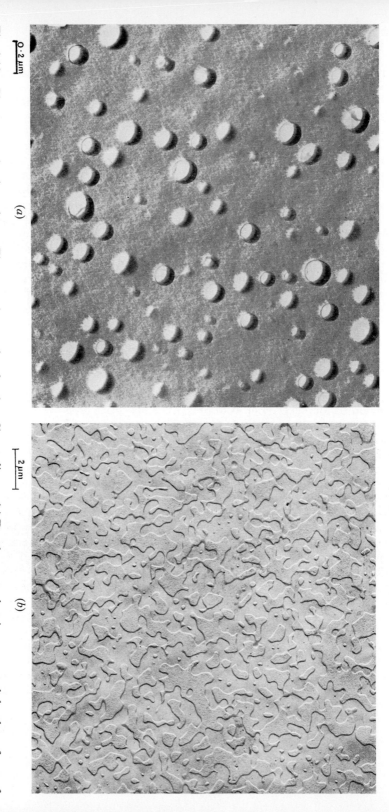

0·2 μm

(a)

2 μm

(b)

Fig. 1.13. Phase separation in a glass. Electron micrographs of carbon film replicas. (a) Droplet type, showing separated droplets of one of the phases ; (b) Vycor type, showing interpenetrating structure of the two phases (C. R. W. Liley).

a sub-microscopic scale glass may not be as homogeneous and perhaps not even as irregular as this model would suggest.

The physical characteristics of glass are certainly consistent with its being a microscopically homogeneous, ionic solid with no long-range order : it is a poor electrical and thermal conductor with no 'free' electrons but shows ionic conductivity at elevated temperatures ; it is isotropic, essentially uniform and transparent to visible light but gives only broad diffuse X-ray or neutron diffraction rings. These diffuse diffraction patterns do not yield *direct* evidence for Zachariason's network theory because, in the absence of long-range periodicity, X-ray or neutron diffraction techniques yield only the average spacings and coordination numbers of atoms or ions which exist in recurring groups. To obtain even this limited information requires a more careful experimental procedure and a much more involved analysis of the diffraction patterns than is needed to determine the structure of simple crystalline materials. Detailed X-ray diffraction experiments have not yet been undertaken on more than a few simple glass systems. Consequently no direct information is available on the location of many cations which occur in complex glasses. The attempts to deduce the positions and coordination of ions from, for example, the variation of density or of optical properties, etc. with composition do not always lead to consistent answers.

It is possible that the structure and composition vary from one region to the next. For example, in a soda–silica glass we might have very small (sub-microscopic) regions of nearly pure silica interspersed with regions having the composition of a sodium silicate. This would certainly not be at variance with the X-ray results and might help to account for some other observations. Electron microscope pictures of very thin films of some glasses taken at very high magnification in transmission (i.e. the electron beam passes through the film) show a very patchy structure. Typically the 'patches' are of order 10 nm across. The transparency of the glass film to electrons is apparently not uniform but there is no general agreement on the interpretation of such micrographs. (The new 1 MeV electron microscopes will enable thicker sections to be examined and may in the near future provide much less ambiguous information.) It has been known for many years that certain mixtures of molten oxides will separate into two immiscible liquid phases and also that some special glass compositions exhibit quite unambiguous phase separation (fig. 1.13), in some cases on a scale large enough to be detected with an optical microscope. Such separation into small regions of different composition may be the only effective way that all the ions present can be surrounded by the number and kind of nearest neighbours required to minimize the energy of the system.

There is, as yet, no general agreement on the relevance of this kind

24

of evidence and argument to the structure of *normal oxide glasses*. The random network theory does provide a very useful qualitative model for the structure and we shall use it in this way in our discussions of physical properties in the following chapters.

Simple models in science can often be justified by the *quantitative* agreement between predictions and experiment. Many can be refined by introducing additional features to allow for initial discrepancies, as for example with the simple kinetic theory of gases and the electron theories of metals. Both the highly-ordered structure of crystalline solids and the entirely random arrangement of molecules in a gas enable straightforward if not always simple mathematical techniques to be used to develop quantitative predictions about their behaviour from models involving the constituent particles and the forces between them—the ultimate objective in physics ! By contrast, glasses, polymers and liquids, which have a locally ordered arrangement of atoms but lack long-range periodicity, are very difficult to treat mathematically ; strong interactions exist between nearby ' particles ' but the detailed arrangement changes with time (liquids) and/or changes from one region to the next. The random network theory has not been developed quantitatively in a way which could provide significant tests of this model for the structure of glass.

CHAPTER 2
the thermal properties of glass

2.1. *Introduction*

MOST of the physical properties of solids can now be understood in terms of the properties of the constituent atoms and the forces between them. For *crystalline solids* it is possible to calculate values for many of the macroscopic parameters which represent their behaviour (that is for things like the expansivity, the elastic constants, the specific heat capacity, etc.) and agreement between the calculated and the experimentally measured values provides the real evidence for claiming that the particular property can be ' understood '. The physical processes which occur at atomic level are essentially the same in glasses and in crystals although, in general, theoretical techniques have not been developed which enable all the macroscopic properties of a large irregular assembly such as is envisaged for glass to be calculated even though the contributing atomic scale behaviour is known.

When explaining or deducing the properties of crystals it is traditional (as well as expedient) to use the simplest model which is consistent with the known behaviour of atoms and will reproduce satisfactorily the particular characteristic of interest. For example, the thermal properties of a crystal can be described quite well using a model in which the atoms are represented by point masses coupled to one another by springs. The mass of each point and the characteristics of the ' springs ' must be chosen to correspond to the atoms and interatomic forces in the crystal of interest. The detailed interactions between atoms (or indeed, the structure of isolated atoms) cannot be adequately represented using Newtonian laws of mechanics, but require the concepts of wave- or quantum-mechanics. It is not surprising therefore to find that even with very simple models it is often necessary to incorporate quantum ideas in order to obtain agreement with experiment. We shall describe the masses and springs models for thermal properties and discuss the effects of quantization in section 2.3. Before we start these more detailed arguments it will be useful to develop an important distinction which divides the properties of a solid into two separate groups.

Experiment shows that the parameters which characterize some, *but not all*, of the physical properties of a crystalline solid are very dependent on the purity of the sample and also on whether it is a single crystal or a polycrystalline aggregate. Evidently some properties are especially sensitive to occasional irregularities in the ordered arrangement of atoms while others are not : there are *structure sensitive* and *structure insensitive* properties. The structure sensitive properties include the electrical and thermal conductivities, the breaking strength or yield point and all the ' loss properties ' such as dielectric loss, the attenuation of acoustic waves, ferromagnetic loss, etc. The structure insensitive properties are those that depend primarily on the nature of the atoms and on the number of and interaction between nearest neighbours ; they include the density, the elastic constants, specific heat capacity and dielectric permittivity. In the theories of crystals a *perfect crystal* model is often used ; similar atoms are all in exactly equivalent positions in an appropriate lattice, i.e. a three-dimensional periodic array. Real crystals are never perfect : in addition to crystal boundaries and impurity atoms there are misplaced atoms, atoms missing from some of the normal lattice positions and whole sequences of displaced atoms which give rise to faults along continuous lines or over planes within the crystal. These imperfections have a marked and often dominating effect on the structure sensitive properties. Perfect crystal models are adequate only for the structure insensitive properties and the study of crystals over the last 30 years or so has been much concerned with identifying the various lattice imperfections and evaluating their separate or collective roles in determining the different structure sensitive properties. Structure insensitive properties are to a large extent independent of both the long-range order and the local faults in the structure of a solid ; these properties are much the same for a crystal and a glass of similar composition. By contrast the structure sensitive properties of a glass are markedly different from those of crystals and the theoretical position is much more complex ; it is no longer possible to relate the large scale behaviour to the properties of specific, isolated, faults in an otherwise regular structure.

2.2. *Additive relations for properties of glass*

Many of the structure insensitive properties of oxide glasses can be represented by simple expressions, almost as though glass were a simple mixture of components and each contributed independently to the overall effect according to the amount present. The so-called *additive relations* take the form

$$A = C_1 X_1 + C_2 X_2 + C_3 X_3 + \dots = \sum_i C_i X_i, \qquad (2.1)$$

where A is the structure insensitive parameter such as the specific volume, specific heat capacity or thermal expansivity, etc., C_i are the weight or molecular fractions of each of the oxide components and X_i are experimentally determined ' factors ' representing the contribution of each component. The factors proposed by A. Winkelmann and O. Schott for the calculation of the specific heat capacity, specific

27

volume and the thermal expansivity are given in table 2.1. More refined expressions have been devised which include an additional term so that

$$A = K + \sum C_i X_i \qquad (2.2)$$

and K is adjusted to represent glasses of different type and glasses cooled at different rates. Alternatively, different sets of factors have been determined for various groups of glasses so that the set most suited to the particular group of interest may be selected. Using these more refined methods, the calculated values are often within a few per cent of the experimentally measured values.

Physical parameter is given by $\sum C_i X_i$, where C_i is the weight fraction of each oxide and values of X_i for linear thermal expansivity, or specific heat capacity or specific volume are given by

Component	Linear exp. $\times 10^6/K^{-1}$	Specific heat cap. $\times 10^{-2}/J\,kg^{-1}\,K^{-1}$	Specific vol. $\times 10^4/m^3\,kg^{-1}$
SiO_2	2·7	8·02	4·35
B_2O_3	0·3	5·33	5·26
Na_2O	33·3	11·2	3·85
K_2O	28·3	7·81	3·57
MgO	0·3	10·2	2·63
CaO	16·7	7·98	3·03
BaO	10·0	2·81	1·43
PbO	10·0	2·14	1·04
Al_2O_3	16·7	8·69	2·44

Table 2.1. Examples of additive factors for the properties of mixed-oxide glasses.

While these empirical relations are useful to the glass technologist for the selection or adjustment of compositions to meet a particular physical specification, they tell us virtually nothing about the fundamental origin of the behaviour. Why *do* glasses (or crystals for that matter) expand when heated ? What determines how much heat energy is needed to raise the temperature of a given solid by a fixed amount and why does it vary from one solid to another ? These are the sort of questions we can answer in a general way by reference to simple models to be described in the following sections.

2.3. Specific heat capacity of solids and atomic oscillations

It is now recognized that the thermal properties of solids, the specific heat capacity, thermal conductivity and expansivity are all related to the vibration of the atoms in the solid. As the temperature is raised the amplitude of vibration and thus the energy associated with each vibrating atom increases. For most solids near room temperature this is the only mechanism leading to a significant change in energy content with temperature so that the heat capacity (specific heat capacity × mass) of these solids is simply the rate of change of the total vibrational energy with temperature. When a temperature gradient exists in a solid the amplitude of the atomic vibrations will be greater in the hotter region. In non-metallic solids, the conduction of heat is entirely due to the transmission of these vibrations, i.e. to the propagation of mechanical waves through the solid.

Model for atomic vibrations

We can start with a very simple model to represent the atomic vibrations. Imagine one of the interior atoms in the solid to be displaced slightly from its equilibrium position; this will upset the normal balance of the attractive and repulsive forces between this atom and its neighbours and will result in a net force on the atom tending to return it to the equilibrium position. We assume as a first approximation that the restoring force is proportional to the displacement and that it is as though each atom were bound to a particular site in the solid. If x is the displacement of the atom, m its mass and β is the restoring force per unit displacement, the force on the displaced atom is $-\beta x$ and the equation of motion (*mass × acceleration = force*) is

$$m \frac{d^2x}{dt^2} = -\beta x. \qquad (2.3)$$

This has a solution

$$x = A \cos [\sqrt{(\beta/m)}]t, \qquad (2.4)$$

where the constant A is the amplitude, and the atom executes a simple harmonic oscillation about its normal equilibrium position. The frequency of vibration $\nu = (1/2\pi)\sqrt{(\beta/m)}$ is determined by the restoring force constant, β, and the mass of the atom. β can be found if the forces of interaction between adjacent atoms are known or for simple solids we can obtain an estimate from the elastic constant.

Let us consider a simple solid in which all the atoms are of the same kind, with average interatomic separation d. Young's modulus Y is

$$\frac{\text{tensile stress}}{\text{tensile strain}} = \frac{(\text{longitudinal force/cross-section area})}{(\text{increase in length/original length})}.$$

If the solid is of length L and cross-sectional area A there will be $\sim L/d$ atoms in the length and $\sim A/d^2$ atoms in the cross section. Suppose we apply a tensile force, F, such that the average longitudinal interatomic spacing increases by x, then

$$F = (A/d^2)\beta x$$

29

and the total increase in length will be $(L/d)x$.   Hence the tensile stress F/A will be $\sim \beta x/d^2$, the tensile strain $\sim x/d$ and therefore

$$Y \sim \beta/d. \tag{2.5}$$

Since typically $Y \sim 10^{11}$ N m^{-2} and $d \sim 2 \times 10^{-10}$ m, β is ~ 20 N m^{-1}. For $m \sim 10^{-26}$ kg the vibration frequency will be of the order of 10^{13} Hz.

Principle of equipartition

In order to calculate the heat capacity of a solid we need to know not only the total number of such atomic oscillators but also their energy as a function of temperature.   The mechanical model itself does not help directly here, but there is a general principle which follows from the laws of classical physics which will enable us to write down an average value for the energy of each oscillator quite simply.

One situation in which this principle is used may be familiar already. When we use the kinetic theory to construct a model of an ideal gas, we discover that on average the energy associated with each particle of a monatomic gas is $3 kT/2$, where k is the Boltzmann constant $(1\cdot38 \times 10^{-23}$ J K$^{-1})$ and T is the absolute temperature.   Thus, according to this theory, the average energy of each particle in the gas increases with temperature.   For an ideal gas there is no interaction between the particles (except during a collision) so that the energy of each particle is entirely kinetic.   The interesting point is that the average value of this energy does not depend on the mass of the particle ; evidently the average speed of the particles at a given temperature is smaller for a gas of greater atomic weight.

We can write the kinetic energy of any one particle in the gas as

$$E = \tfrac{1}{2}mv^2 = \tfrac{1}{2}mv_x{}^2 + \tfrac{1}{2}mv_y{}^2 + \tfrac{1}{2}mv_z{}^2, \tag{2.6}$$

where v is the instantaneous velocity of the particle, v_x, v_y, v_z are the cartesian components of this velocity and m is the mass of the particle. The average energy for all the particles will therefore be given by

$$\bar{E} = \tfrac{1}{2}m\overline{v^2} = \tfrac{1}{2}m(\overline{v_x{}^2 + v_y{}^2 + v_z{}^2}).$$

Since v_x, v_y and v_z are independent this can be written

$$E = \tfrac{1}{2}m(\overline{v_x{}^2} + \overline{v_y{}^2} + \overline{v_z{}^2}).$$

If there is no net flow of gas particles (i.e. the centre of mass of the gas is at rest), the values of $v_x{}^2$, $v_y{}^2$ and $v_z{}^2$ averaged over all particles or over a long time for any one particle must be equal.   Hence

$$\bar{E} = \tfrac{3}{2}m\overline{v_x{}^2} = \tfrac{3}{2}m\overline{v_y{}^2} = \tfrac{3}{2}m\overline{v_z{}^2}$$

and we can also write

$$\tfrac{1}{2}m\overline{v_x{}^2} = \tfrac{1}{2}m\overline{v_y{}^2} = \tfrac{1}{2}m\overline{v_z{}^2} = \tfrac{1}{2}kT. \tag{2.7}$$

This is a particular example of a general principle which applies to potential energies as well as to kinetic energies and is known as *the principle of equipartition*. Formally this can be expressed in the following way : the average energy of any particle is equivalent to $\frac{1}{2}kT$ for each independent coordinate required to specify its instantaneous energy. We can regard the average to be taken either over a period of time large compared with the time between collisions for one particle, or over many particles at one instant in time.

Calculation of specific heat capacity

The principle of equipartition will also give the average energy associated with each of the atomic oscillators in our solid and we can, therefore, knowing the total number of oscillators, calculate the total thermal energy and hence the heat capacity. Each atom in the solid can vibrate independently in three mutually perpendicular directions and each of these vibrations will have both potential and kinetic energy associated with it at any arbitrary instant. It is necessary therefore to use a velocity and a position coordinate in each of three perpendicular directions, giving a total of six independent coordinates, in order to write down a general expression for the instantaneous energy of a given atom. The average thermal energy of each atom will, therefore, be $6(kT/2)$ and hence the thermal energy in the solid will be $3nkT$, where n is the number of atoms, or $3N_A kT$ J mol^{-1}, where N_A is the Avogadro number $(6 \cdot 03 \times 10^{26}$ mol$^{-1})$.

Thus the heat capacity per mole will be

$$\mathrm{d}/\mathrm{d}T\,(3N_A kT) = 3N_A k \sim 2\cdot5 \times 10^4 \text{ J mol}^{-1}\text{ K}^{-1},$$

which does *not* depend on the characteristics β and m of the oscillators or on the temperature. According to this model the heat capacity per mole of atoms should be the same for all solids and should be independent of temperature. Experimental observation shows that for many solids at elevated temperature this is indeed correct ; it is embodied in the empirical law of Dulong and Petit. At lower temperatures however the molar heat capacity falls below $3N_A k$, the temperature at which this first occurs varying from one solid to another (fig. 2.1).

This decrease in the heat capacity with temperature cannot be explained using the concepts of classical physics. The significant error in the model outlined above is the use of the principle of equipartition. Einstein first realized that the revolutionary idea of quantization of energy, developed by Planck in order to account for the distribution of energy in the spectrum of the radiation emitted by a black-body, could be applied to the atomic oscillators in a solid. Adopting Planck's hypothesis that energy is quantized, Einstein assumed that only certain discrete energies (and thus amplitudes) of vibration are possible and that the energies available to an atomic oscillator are related to the

31

frequency of oscillation by $E = n\,h\nu$, where n is an integer and h is the Planck constant (6.63×10^{-34} J s). If the energy is quantized then the principle of equipartition is not valid at low temperatures : the discontinuities in the energy change drastically the expression for the average energy per oscillator.

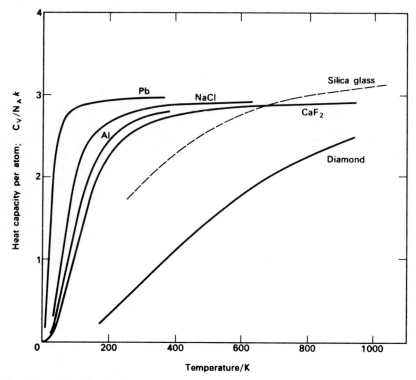

Fig. 2.1. Variation of heat capacity per atom, $C_V/N_A k$, with temperature for various solids.

Einstein showed that the average energy of quantized oscillators of frequency ν is given by

$$\bar{E} = h\nu/(\exp\,(Z) - 1), \qquad (2.8)$$

where $Z = h\nu/kT$. Note that for $kT \gg h\nu$, $\exp\,(Z) = 1 + h\nu/kT$ and thus $\bar{E} \simeq kT$. That is, at 'high temperatures' this expression gives the same result as the principle of equipartition but when $T \gtrsim h\nu/k$ the average energy of quantized oscillators is smaller than kT.

We can now use this expression for the average energy per oscillator to calculate the thermal energy and hence the heat capacity of our solid. We have $3N_A$ oscillators of frequency $\nu = \sqrt{(\beta/m)}$ corresponding to the

32

three independent directions of vibration for each of N atoms so that the total energy will be

$$3N_A \bar{E} = 3N_A h\nu/(\exp{(Z)} - 1) \text{ per mole} \qquad (2.9)$$

and the molar heat capacity, which is just the rate of change of energy with temperature

$$\frac{\mathrm{d}(3N_A \bar{E})}{\mathrm{d}T} = 3N_A k \left(\frac{Z^2 e^{-Z}}{(1 - e^{-Z})^2} \right).$$

This can be written as

$$3N_A k f(Z). \qquad (2.10)$$

$f(Z)$ is known as the Einstein function ; it varies from 0 to 1 as T goes from 0 to $\gg (h\nu/k)$ K.

We can see in a qualitative way why quantization of the energies of the oscillators has such a marked effect on the heat capacity at low temperatures. When the temperature is low, $kT \ll h\nu$, a small increase in the temperature causes only a very small fraction ($\sim \exp{-h\nu/kT}$) of the total number of independent oscillators to increase their energy to that of the next available level and therefore not much energy is absorbed from the heat source. At higher temperatures where $kT > h\nu$, all the oscillators will increase their energy of vibration as the temperature increases, the spacing between the discrete levels is small compared with the average thermal energy and the behaviour approaches the classical form.

We have so far been dealing with a solid in which all the atoms are of the same kind so that the mass and the spring constant are the same for each oscillator. In a typical glass there will be several different kinds of ion, having different values of m and of β and thus of oscillation frequency. The ions such as Si^{4+} and O^{2-} which are strongly bound into the network will have higher values of β, and therefore of frequency, than the loosely held, network modifying, alkali metal ions. In summing up the contributions from the independent oscillators according to the Einstein model, for a glass we must allow that different ions will have different characteristic frequencies and hence also, at a given temperature, different values of the Einstein functions. If unit mass of the glass contains $n_1, n_2, n_3 \ldots$ atoms of each kind with different m and/or β then the specific heat capacity will be

$$C_v = 3n_1 k f(Z_1) + 3n_2 k f(Z_2) + 3n_3 k f(Z_3) + \ldots, \qquad (2.11)$$

where $Z_{1, 2, 3}$ are $h\nu_{1, 2, 3}/kT$. This is an equation of similar character to the empirical additive equations described in section 2.2. Approximate comparisons with the empirical Winkelmann factors suggest that the Si^{4+} and O^{2-} ions of the network each contribute at room temperature only about 0·6 of the full classical contribution ($3k$) while the

33

alkali metal ions and the non-bridging oxygens are virtually classical oscillators each contributing $3k$.

For a simple crystalline element all the atoms are identical and, according to the Einstein model, have the same vibration frequency (the surface atoms and other imperfections normally form a negligible fraction of the total and are ignored). The variation of the specific heat capacity with temperature should follow the Einstein function $f(Z)$. At very low temperatures as $T \rightarrow 0$ K there are small but significant discrepancies between the predictions of the Einstein model and the observed results for crystals. In recent theories of crystal specific heat capacity the approximation that each atom or ion is bound to a fixed point in the lattice is abandoned ; it is recognized

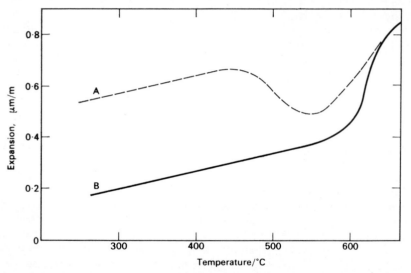

Fig. 2.2. Thermal expansion of a chilled (high fictive) sample (A) and an annealed (low fictive) sample B of the same glass.

that the atoms are bound to one another, and that a wide spectrum of vibration frequencies is possible corresponding to the many coupled oscillations of the atoms on the lattice. Calculation of the frequency spectrum is a somewhat formidable task even for quite simple crystals and comparable analyses have not been attempted for glasses. However, the Einstein model provides a useful semi-quantitative picture at normal temperatures.

At very high temperatures there are further complications : as the temperature approaches the transformation range for the glass, increases in the amplitude of vibration of the ions are no longer the only mechanism leading to absorption of energy with increase of temperature.

In this temperature range changes in the molecular arrangement can occur which can absorb energy and also cause changes in the specific volume. The exact behaviour of a given sample of glass depends on its previous ' thermal history ' in this temperature range. In samples with high fictive temperature the molecular arrangement will change to a more compact form, reducing the fictive temperature, if the actual temperature of the sample is increased very slowly into the lower end of the transformation range. While such changes to a configuration corresponding to a lower fictive temperature are taking place there may be a transitory *evolution* of heat and a volume contraction, even though the actual temperature of the sample is increasing (figs. 2.2 and 2.3).

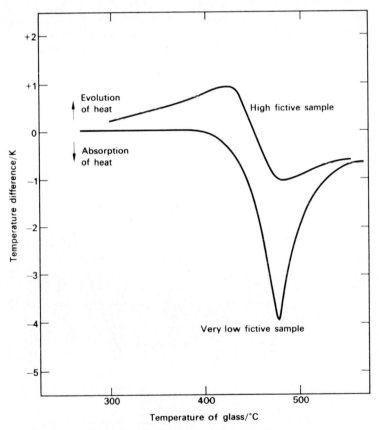

Fig. 2.3. Heat evolution and absorption effects in the transformation range : the temperatures of two samples of the same glass slowly heated in a furnace are compared with the temperature of a neutral body in the furnace.

35

2.4. *Thermal expansion*

The model of a solid as an array of oscillating masses also provides a simple explanation of why it is that solids normally expand on heating. However, we must not be tempted into equating thermal expansion to the increase in amplitude of vibration of the atoms as the temperature increases. For many crystals the amplitude of atomic vibrations at the melting point is about one-eighth of the interatomic spacing and typically the change in interatomic spacing due to thermal expansion between 0 K and the melting point is $\leqslant 1$ per cent. Thus the *total* expansion is only ~ 8 per cent of the maximum amplitude of vibration. If the oscillations were strictly harmonic, i.e. the restoring force proportional to the displacement, then the average distance between atoms would not change with the amplitude of vibration (see fig. 2.4). The interatomic spacings, and therefore the dimensions of a solid, change with temperature only because the oscillations are not harmonic.

Let us consider for a moment the force between two oppositely charged ions which are remote from all other material. When the ions are far

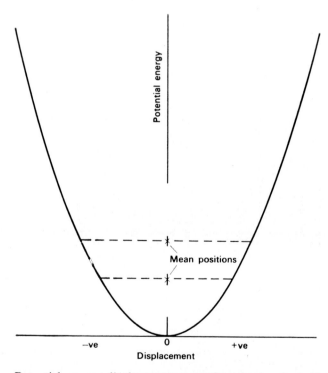

Fig. 2.4.   Potential energy–displacement curve for a harmonic oscillator ; the average displacement of the oscillator is zero for all values of the total energy.

36

apart the opposite charges will result in an electrostatic attractive force which increases as the inverse square of their separation. When the separation becomes comparable with the diameters of the ions an additional *repulsive* force appears, due to an interaction between the electrons in the closed shells of the ions. As the separation decreases, this repulsive force increases much more rapidly than the electrostatic attractive force and in principle the ions would take up an equilibrium separation where the repulsive and attractive forces balanced (fig. 2.5). With the ions at this separation it would require a somewhat greater force to decrease their separation by a given distance than to increase it by the same distance. In a mass and spring model of this interaction we should need a spring connecting the two masses which is slightly easier to stretch than to compress.

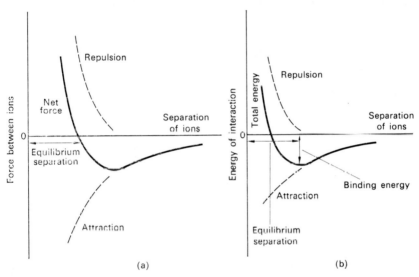

Fig. 2.5. Interaction between a pair of oppositely charged ions. (*a*) Force–separation curves ; (*b*) energy–separation curves.

The asymmetry in the force between the ions can be represented mathematically by writing the net force as

$$F = -\beta x + \gamma x^2 + \text{terms in higher powers of } x \qquad (2.12)$$

or the potential energy as

$$V = \int F \mathrm{d}x = -\tfrac{1}{2}\beta x^2 + \tfrac{1}{3}\gamma x^3 + \dots, \qquad (2.13)$$

where $x = (d - d_0)$ is the increase in the interionic distance from the equilibrium separation.

This kind of asymmetry in the force between atoms is quite general ; it occurs for all types of attractive force—ionic, covalent, metallic and van der Waals' forces. Moreover the same kind of behaviour will be exhibited by an isolated pair of atoms and by the whole group of atoms forming a solid ; the potential energy of an atom or ion in the solid will vary with the interatomic separation like equation (2.13) and fig. 2.6.

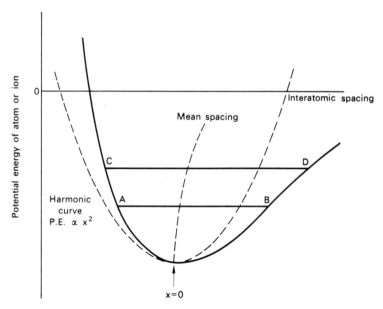

Fig. 2.6.   Potential energy of an atom in a solid as a function of the interatomic spacing.

In terms of d the interatomic or interionic separation, the potential energy curve can be represented approximately by an equation of the form

$$V = A/d^m - B/d^n,$$

where A and B are constants.   The first term, with, $m = 11$ or 12, represents the energy of the repulsive interaction and the second term represents the energy due to the attraction.   The value of n depends on the nature of the dominant attractive forces, e.g., $n = 1$ for ionic bonding and $n \simeq 6$ for van der Waals attraction.   Equation (2.13) can be derived from this expression, using the fact that $dV/dx = 0$, when $x = d - d_0 = 0$, and some algebraic manipulation.   The shape of this curve implies (compare equation (2.12)) that the externally applied force required to compress or to stretch a solid varies non-linearly with the displacement.   However, the departure from linearity is too small

38

to be detected experimentally on normal samples; Hooke's law *is* obeyed, and the asymmetry is too small to be revealed as a difference in the elastic constants for compression and extension. For ordinary samples of most materials *plastic deformation*, which involves rearrangement of the atoms, or *brittle fracture* occurs long before the elastic strain, the fractional change in atomic separation, is as large as 1 per cent. Nevertheless the departure from linearity is very significant for thermal vibrations of the atoms because locally the fractional change in spacing can exceed 10 per cent.

The anharmonic terms, of which γx^2 in equation (2.12) is usually the most important, have a very small effect on the heat capacity of the solid so that we could ignore them in the previous section without significant error, but they are essential for the explanation of thermal expansion (and of the thermal conductivity of non-metallic crystals).

We shall not develop in detail a quantitative model for the thermal expansion of a solid but simply follow a physical argument which will illustrate the nature of the effect of anharmonicity.

If the potential energy of an atom as a function of position is of the form in equation (2.13), the cubic term will result in an asymmetry about the position of minimum potential energy (at $x = 0$) which corresponds to the equilibrium position of the atom. The curve is illustrated in fig. 2.6. At the extreme displacements the vibrational energy of the vibrating atom is, instantaneously, entirely in the form of potential energy ; the kinetic energy is zero. We can represent the total energy of an atom vibrating in a potential well by a horizontal line, such as AB or CD, in fig. 2.6 and then the kinetic energy at any point is given by the vertical distance between the horizontal line and the curve representing the potential energy.

As the temperature is raised the energy of the atom increases and this corresponds to an increase in the amplitude of vibration. An atom vibrating between the positions corresponding to A and B in fig. 2.6 will, following an increase in temperature, acquire additional energy and vibrate between positions such as C and D. Because of the asymmetry in the potential energy curve, the *mean* position of the atom will be displaced as the amplitude of vibration increases from AB to CD. It is this displacement of the mean positions of the atoms in a solid which constitutes thermal expansion.

The potential energy curve for an atom or ion in a particular solid provides, in principle, a great deal of information about the properties of the solid and we shall refer to it on several occasions in later chapters. It is very difficult to calculate the exact form of the curve for real solids but it is possible to use the general shape to explain qualitatively the connection between certain physical properties of materials. For example it is generally true that the higher the melting point of a material the lower is its thermal expansivity at room temperature, table 2.2.

Glass	Softening point/(°C)	Linear exp. $\times 10^7/K^{-1}$	Polycrystalline	Melting Pt./(°C)	Linear exp. $\times 10^7/K^{-1}$
Silica	1600	5·5	Tungsten	3380	43
96% Silica	1500	8	Platinum	1770	93
Aluminosilicate	940	33	Nickel	1453	130
Borosilicate (Pyrex)	820	32	Copper	1084	168
Soda–lime–silica	700	92	Silver	961	180
Lead silicate	580	91	Aluminium	660	230
High lead	380	100	Cadmium	321	315
			Sodium	98	700
			Diamond	>3500	12
			Magnesium oxide	2900	100
			Aluminium (Al_2O_3)	2050	87
			Quartz	1710	80 ∥ axis
					134 ⊥ axis
			Fluorspar (CaF_2)	1360	190
			Sodium chloride	800	400

Table 2.2. Thermal expansivities and melting points of crystalline solids, or softening points of glasses.

In a general way we can see why this should be so by the following argument.

The depth of the minimum in the potential energy curve gives the *binding energy*, or latent heat of sublimation, for an atom of the solid ; the melting point of many crystals is given approximately by $kT_m \sim \frac{1}{2}$ (binding energy) so that a high melting point implies a deep minimum in the curve. The thermal expansivity is determined by the asymmetry in the potential energy curve, and the deeper the minimum the more symmetrical the curve near the bottom, fig. 2.7. At room temperature the average energy of vibration of all atoms will be roughly the same no matter what solid they are in ; thus the lines representing the energy and amplitude of the atomic vibrations on the potential energy curves will be the same distance from the bottom of the curve whatever its shape. (For the one-dimensional vibrations represented in fig. 2.7, this distance is $\sim kT$.) Therefore a higher melting point implies a deeper minimum of higher symmetry and this results in a smaller thermal expansion.

As we have already indicated in Chapter 1 (see fig. 1.1) the thermal expansion of a glass below its transformation range is much the same as that of the crystal of similar composition. Evidently the potential energy curve depends more on the nature and local arrangement of nearest-neighbour atoms, both of which are similar in the glassy and crystalline form of the same substance, than on the exact positions of the more distant atoms.

Silica glass is one of the materials which shows anomalous thermal expansion behaviour : the expansion coefficient for this glass is very much lower than for quartz, table 2.2, and it becomes negative below about $-80°C$. This unusual behaviour, in common with several other anomalous properties of silica glass (see section 5.3 on anomalous elastic behaviour) is not yet fully understood ; it may be related to the very open structure of the network and the consequent predominance of vibrational modes involving displacements of the silicon and oxygen ions transverse to the bond direction.

2.5. *Thermal conductivity*

Thermal conductivity is a structure sensitive property and the barest outline of the physical processes involved requires a more sophisticated picture of a solid than that of independent masses oscillating as if fixed to separate points in space. We now must recognize that the atoms are in fact bound to one another and that atomic thermal vibrations are essentially standing waves in the solid. There will be a range of possible frequencies of the standing waves in much the same way as for standing waves on a length of string. There are two important differences for waves in the atomic solid : the energy of the waves is quantized, again $E = nh\nu$, and there is an upper limit to the possible

41

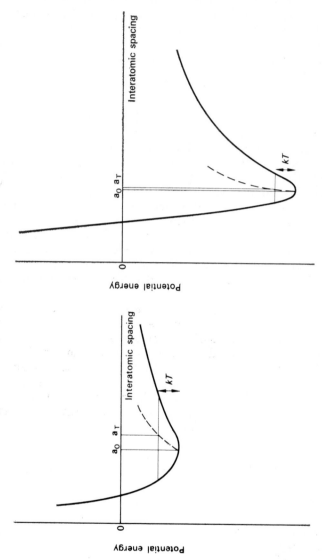

Fig. 2.7. Thermal expansion depends on the depth of the minimum in the potential energy curve ; a_0 corresponds to the equilibrium spacing at 0 K and a_T to the mean spacing at T K.

42

frequency, a lower limit to the wavelength. This limitation arises because it is not possible to produce a wave in a lattice of discrete particles with a wavelength shorter than twice the interparticle spacing. For a *continuous* string the wave displacement has a real value at any point along the string but for a linear array of particles the wave displacement can only be defined at the positions occupied by particles, see fig. 2.8. The particles lying on a mathematical waveform with $\lambda < 2a$ have exactly the same positions as those on another wave with $\lambda > 2a$, moreover there is *nothing* which *can* be displaced to correspond with all the peaks of the shorter wave. The more sophisticated theories of the specific heat of crystals are based on this model and yield results in better agreement with experiment at very low temperatures.

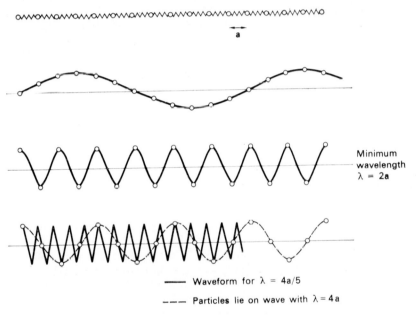

Fig. 2.8. Waves on a linear lattice.

The conduction of heat in a non-metallic solid is due to the propagation of these mechanical waves through the material. In fact the longitudinal waves are just high frequency ' sound ' waves and they will propagate with the speed of sound in the material. However, everyday experience shows that it takes quite a long time for heat energy to be conducted from one part of a solid to another. An analogous situation occurs in gases where the average speed of a gas molecule is very much larger than the rate at which energy (or matter, in the interdiffusion of two gases) can be transported from one region to another. In both cases the paradox is resolved by admitting that the progress of a wave or a molecule respectively in any given direction is continually interrupted by collision. In a gas, molecules are frequently in collision and a 'hot' molecule or a 'different' molecule is scattered into a new random direction after travelling a very short distance, about 10^{-7} m at normal pressures. In a crystal, collision between lattice waves can occur *because* the atomic oscillations are anharmonic. If the atomic oscillations which make up the lattice wave were strictly harmonic, then one wave pulse could pass straight through another and there would be no interaction other than the local and temporary change in amplitudes in the region of overlap. The existence of anharmonic terms means that the presence of one wave pulse in a particular region will alter slightly the properties of that region, so that it presents to a second pulse effectively a 'stiffer' patch in the crystal which can cause scattering.

We can borrow ideas from the kinetic theory of gases to give a simple account of lattice conductivity. For a gas the thermal conductivity according to elementary kinetic theory is given by

$$\kappa = \rho c_v v \lambda / 3, \tag{2.14}$$

where ρ is the density,
   c_v is specific heat capacity at constant volume
   v is the average molecular speed
and λ is the mean free path between collisions.
This can be written

$$\kappa = C v \lambda / 3,$$

where C is the heat capacity per unit volume $= \rho c_v$. These expressions follow from the random motion of the gas particles, each of which has an energy proportional to c_v and an average speed v between collisions, distance λ apart. We can picture the conduction of heat by a solid simply by regarding the solid as a box containing wave pulses which can collide with one another and with obstacles in the box. The obstacles represent the effect of lattice defects in a crystal. By analogy with the kinetic theory of gases we can write

$$\kappa_{\text{solid}} = C' v_0 \lambda_w / 3, \tag{2.15}$$

C' will be the heat capacity per unit volume and at high temperatures $C' = \rho 3 N_A k$; v_0 will be the speed of sound in the crystal and λ_w the mean free path of the lattice waves.

Intuitively we can expect that λ_w will vary inversely as the density of lattice waves just as λ does with gas density. At low temperatures C' falls because waves of high frequency, like the oscillators in the Einstein model for heat capacity, are not excited. Nevertheless the

44

thermal conductivity of a crystalline solid may increase at low temperature because the increase in λ_w more than compensates for the loss of some of the waves which can transport energy. Impurity atoms and other imperfections in a crystal will also scatter the lattice waves and will determine the upper limit to the mean free path since their concentration is not reduced as the temperature falls. As we should expect from these arguments and equation (2.15), the thermal conductivity of glasses is much lower than that of crystals : the specific heat capacities of crystalline and glassy materials are comparable but in the highly disordered glassy structures the mean free path of lattice waves will be very short and not sensitive to changes in temperature. Figure 2.9 illustrates the difference between crystals and glasses ; the conductivity of glasses decreases at low temperatures because the number of waves falls and there is no compensating increase in the mean free path, since this is fixed by the disorder in the structure.

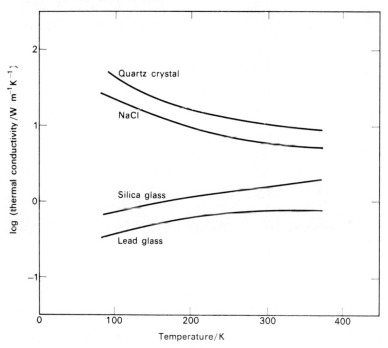

Fig. 2.9. Thermal conductivity versus temperature for some crystals and glasses.

2.6. Thermal properties of glass—practical considerations

The experimental methods used for the routine measurement of the thermal properties of glass are usually refined versions of the classical

45

methods familiar in most school laboratories. For example, specific heat capacities are measured by the method of mixtures, thermal conductivity by a variant of ' Lee's disc method for a poor conductor ' and so on. The experimental procedures differ in detail rather than in principle from those of traditional introductory physics : the calorimeters will be more sophisticated to reduce heat losses, temperatures will be measured by multi-junction thermocouples and ovens or furnaces used for heating samples to a known temperature will be elaborately designed to ensure uniformity of temperature and to enable accurate control and measurement.

Fig. 2.10. Typical small glass-to-metal seals.

From a technological viewpoint the thermal expansivity is the most important thermal characteristic and for many applications the ability to produce glasses with accurately controlled expansivities is vital. Glass will fuse to a variety of metals and form an impermeable seal. The bond between the metal and the glass is formed by a very thin coherent layer of metal oxide and usually a thin layer of ' metal-oxide glass '. In applications where a metal wire has to be sealed into a glass vessel to form an electrical connection, as in light bulbs and radio valves for example, it is essential that the thermal expansion of the metal and the glass should be closely similar. If there is a difference between the thermal contraction of the two materials as the temperature falls from the softening point of the glass to room temperature, stresses will be set up which can lead to fracture of the glass, see fig. 2.11. Precise matching of the expansion characteristics over the whole range of temperatures up to the softening point becomes essential when metal wires or rods of large diameter are to be sealed into glass. A variety of ' sealing glasses ' has been developed which match closely the expansion of particular metals. Tungsten and molybdenum have very low expansivities and will form good seals with borosilicate glasses ; thin wires can be sealed to Pyrex glass but because the expansion characteristics do not match exactly for wires above ~ 1 mm diameter a special sealing glass is required. Alloys of Ni, Fe, Co, which are easier to work than the very brittle tungsten and molybdenum, have also been developed to match the expansion of various borosilicate glasses. Most of the very fine ' lead in ' wires for small lamps and bulbs are more economically provided by a Ni/Fe alloy sheathed in copper sealed into a high electrical resistivity, lead–alkali–silica glass

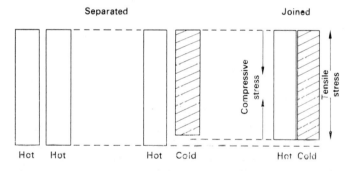

Fig. 2.11. Thermally induced stresses.

which will readily join to the soda–lime–silica bulb. Conflict between the desired expansion characteristics of the sealing glass and that of the main bulb or envelope of a large vacuum tube can be resolved by using a succession of different glasses to produce a ' graded seal ' ; the expansion difference between adjacent glasses is limited to that which will not lead to intolerable stresses. A similar procedure can be used to join small glass tubes of widely different expansion coefficient, e.g. a silica tube to one of soda–lime–silica glass, through a series of short tubular sections of ' intermediate ' glasses.

Where glass is to be sealed to the edge of a metal sheet or disc or to the end of a metal tube and a very soft metal such as copper can be used, glass–metal seals can be produced without ' matching ' of the expansivities. Provided the expansion of the metal is greater than that of the glass and the edges of the metal sealed to the glass are sufficiently thin, then the differential strains which occur during cooling can be accommodated by *plastic* deformation (yielding) of the metal. This is the Housekeeper seal, named after W. G. Housekeeper ; it is a particularly useful technique for joining glass and metal tubes but is seldom now used for sealing metal lead wires through the wall of a glass vessel. A variety of small, glass–metal seals are illustrated in fig. 2.10.

The tendency for ordinary bottles and jars (i.e. those made from the cheaper soda–lime–silica glasses) to crack as they are immersed in or filled with hot liquids is well known to kitchen helpers and experimenters. The ability to withstand sudden heating or cooling without cracking is called thermal endurance or thermal shock resistance. When one surface of a thick section of glass is suddenly heated or cooled the poor thermal conductivity of the material results in a large temperature gradient, the colder layers prevent free expansion of the hotter ones and stresses, which may be sufficient to crack the glass, develop in the cross section, fig. 2.11. The magnitude of the stresses set up for a given change in temperature depends on the thickness of the glass and the thermal conductivity, as these determine the temperature gradient, and also on the expansivity and the elastic constant, as these determine the ' free thermal strain ' and the stress required to suppress it respectively. The shape of an article is also of particular importance ; sudden changes in thickness or sharp corners can lead to extra high, local stresses. As far as the glass itself is concerned we could reduce the thermal shock stress by increasing the conductivity or the breaking strength, or by decreasing the elastic constant or the thermal expansion. Of these the *only* one which is at all sensitive to the composition of the glass is the thermal expansivity, and in practice high thermal shock resistance is obtained by selecting a glass with a low expansivity. This is why borosilicate glasses such as Pyrex are used to produce ovenware ; aluminosilicate glasses are used for combustion tubes and top-of-the-stove ware (e.g. the glass frying-pan) because these glasses have higher softening points than ordinary borosilicates. By far the best glass available from this point of view (it can be quenched from red heat into water) is silica and this is used for the most exacting conditions.

The transfer of heat through glass is of considerable technological importance in windows of various kinds and also within the glass-manufacturing industry in the design of furnaces for melting or annealing large masses of glass. Thermal conduction is not the only process of heat transfer ; glasses are partially transparent to radiation in the infra-red range ($1-5 \times 10^{-6}$ m) and heat transfer by radiative processes plays an important part in practice.

48

The thermal conductivity, κ, of a material is defined by the equation

$$\frac{\mathrm{d}Q}{\mathrm{d}t} = \kappa A \frac{\mathrm{d}T}{\mathrm{d}x} \qquad (2.16)$$

(rate of heat flow $=$ (thermal conductivity) × (cross-sectional area)
or heat flux) $\quad$ × (temperature gradient)

For a *thin* plate of glass adjacent to a high temperature heat source, the total heat flux through the thickness will be the sum of that transported by conduction (i.e. by the ionic vibrational waves) within the structure plus that radiated through the glass. For a *thick* block of glass the intensity of the radiation transmitted directly through the thickness will be negligible. The transmitted intensity falls off exponentially with thickness—Lambert's law $I_{\mathrm{Trans}} = I_{\mathrm{Incid}} \exp(-\beta_T t)$, where $t =$ thickness and β_T is a constant (see Chapter 4). However, radiation entering a *thick* block from the high temperature source is absorbed in the first layers of the glass and raises the local temperature, whereupon the intensity radiated in turn by these layers is increased. Progressive absorption and re-radiation in the successive layers of the block provides, once a steady state has been reached, an additional flux of heat through the block. The heat transferred by this process becomes comparable with that transferred by the normal conduction process at temperatures of the order of 500°C and increases rapidly at higher temperatures. This is therefore a dominant process in the transfer of heat into glass in large tank-furnaces, but, oddly enough, the significance of the effect has been recognized only quite recently. Calculation of the total heat flux is rather complex and beyond our scope here : a major difficulty arises because the heat flux due to the radiative process not only varies rapidly with the mean temperature but also depends on the size and shape of the glass block.

At room temperature and below, the heat flux is entirely due to the normal conduction process and is satisfactorily represented by equation (2.16) above with κ a property of the material itself and independent of the size and shape of the specimen.

49

CHAPTER 3
the electrical properties of glass

GLASS is used extensively in the manufacture of a very wide range of electrical products, from lamps and electronic valves of all kinds to insulators both large and small and even electronic circuit components. Not only have glasses a desirable range of electrical properties but they are cheap, they can be formed readily into complex shapes and they often have essential additional properties, such as the ability to transmit visible radiation, to seal to metals, and to resist chemical attack.

Direct-current conductivity

3.1. *Introduction*

In any introductory description of the electrical properties of materials, glass is almost certain to be cited as an example of a good electrical insulator. The resistivity of a soda–lime–silica glass at room temperature is of order 10^8–10^9 Ω m, about 10^{17} times greater than that of copper, although at elevated temperatures the resistivity of glass can fall to a few Ω m. Thus the electrical resistance across a tank of molten glass can fall to a few ohms and this makes it possible to heat the glass directly by the passage of an electric current.

The conductivity of glass is due to the mobility of certain ions within the structure : it is an *electrolytic* conductor. In soda glasses, it is the sodium ion jumping from hole to hole through the silica network that gives rise to the flow of current. There are some new semiconducting glassy materials, with a resistivity at room temperature in the range $\sim 10^2$–10^6 Ω m, in which *electronic* conduction occurs but these are not silica-based oxide glasses.

Some of the early experiments which indicated that the electric current in a soda–lime–silica glass was due to the movement of sodium ions through the glass were carried out by R. C. Burt in 1925 using ordinary electric light bulbs. The bulb was partially immersed in a bath of molten sodium nitrate which served as an anode and the stream of thermionically emitted electrons from the heated filament was used to complete a d.c. circuit between the inner and outer surfaces of the bulb (fig. 3.1). Sodium released inside the bulb condensed on the cooler upper surface and the quantity transported through the wall of the bulb by a known quantity of electricity (current × time) could be determined after breaking open the bulb. Burt demonstrated that the amount of sodium deposited in the bulb corresponded with that expected according to Faraday's law and hence that the sodium ions were the only significant mobile carrier. More recently I. Peychès has used a radioactive tracer method to study electrical conduction in soda–lime glasses. By irradiating the glass with neutrons, a fraction of the sodium Na^{23} ions were converted *in situ* to Na^{24} which is radioactive with a half-life of 15 hr. A direct current was passed through the glass at 320°C using $NaNO_3$ electrodes and the rate at which sodium ions from the glass appeared in the cathode bath was ascertained from the increase in the radioactivity of the bath. Again the amount of sodium transferred into the cathode bath corresponded to that expected according to Faraday's law. Peychès also concluded that the current in soda–lime–silica glasses is carried exclusively by the positively charged sodium ions.

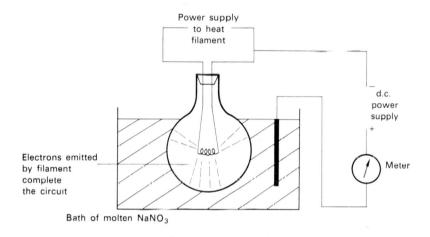

Fig. 3.1. Early method for the measurement of d.c. resistivity of glass using a light bulb.

In these experiments on the electrolysis of glass at elevated temperatures, it is essential to provide a supply of ions at the anode surface to avoid the formation of a layer depleted of positive ions (electrode polarization) as the Na^+ migrate away in the electric field. By using electrode baths of salts other than sodium, e.g. $LiNO_3$ or NH_4NO_3, it is possible to study the extent to which the sodium ions in a given network may be replaced by other ions. It has been shown that it is possible to replace a large fraction (but, significantly, not *all*) of the original sodium ions in a soda–lime–silica glass by ammonium ions or by silver ions but not by potassium or lithium ions. Although potassium and lithium ions will penetrate the glass, they cause the

51

sample to shatter after current has been passing for some time ; the network which was formed originally around the sodium ions becomes highly strained when the smaller Li^+ or larger K^+ ions are substituted. In fact this effect can now be used in a controlled fashion deliberately to develop compressive stresses in the surface of a glass article in order to increase its mechanical strength (see Chapter 6).

3.2. Measurement of resistivity

The d.c. resistivity of silicate glasses ranges from $\sim 10^{15}$ to 10^{-2} Ω m depending on the composition and the temperature. To cover this very wide range, a variety of experimental methods is used. At the lower end the classic Wheatstone bridge or potentiometer methods are adequate. For the higher resistivities, the current is extremely small

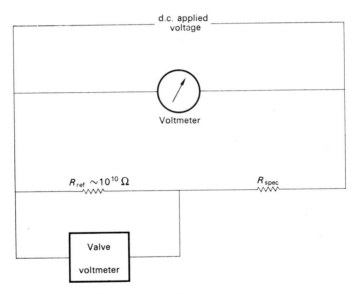

Fig. 3.2. Simple circuit for resistance measurement.

even for very thin discs of glass and these methods become hopelessly insensitive. High resistances are often found by measuring the voltage across the sample and the current passing through it, although special techniques are used to measure currents ($\sim 10^{-9}$ A) which are below the range of commercial galvanometers. By adding a very high resistance (10^9–10^{11} Ω) of known value in series with the sample, the current can be found by measuring the voltage across this series resistance (fig. 3.2) using a valve voltmeter or an electrometer valve in order to avoid shorting it out.

52

In all measurements of the volume resistivity of glasses particular care must be taken to avoid spurious results due to (i) the surface conductivity, (ii) an ' anomalous ', transient, volume conductivity and (iii) the polarization at the electrodes. The current in a thin surface layer on glass can be larger under some conditions than the current through the bulk of the glass. This surface conductivity results largely from the adsorption of moisture from the atmosphere. Water adsorbed on the surface of a soda–lime glass can leach out mobile alkali metal ions near the surface, by exchanging H^+ ions for the alkali metal, and form a thin layer of liquid electrolyte on the surface. Surface conduction increases with the relative humidity of the atmosphere and with the mobility and concentration of the alkali ions in the glass ; it also varies with the previous history of the surface.

For many glasses, particularly at low temperatures, the initial current which flows immediately after the application of an electric field is much larger than the eventual steady state current. This ' anomalous ' conductivity can persist for many minutes and even hours. We shall discuss the nature and origin of the effect later in the chapter since it is related to the processes which contribute to dielectric losses. The existence of this transient conductivity does complicate the measurement of resistance, since polarization of the electrodes cannot be avoided by using low frequency a.c. voltages. The a.c. conductivity of a glass is often significantly greater than the true d.c. conductivity because the former includes a contribution from the anomalous current.

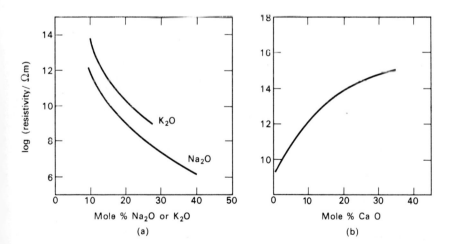

Fig. 3.3. Effect of composition on resistivity at 25°C. (a) For glasses in the series : x mole % K_2O (or Na_2O), 20 mole % PbO, $(80-x)$ mole % SiO_2 ; (b) for glasses in the series : 18 mole % Na_2O, x mole % CaO, $(82-x)$ mole % SiO_2.

At higher temperatures, above about 300°C for most silicate glasses, the anomalous conductivity is very small ; more correctly, perhaps, the transient effects occur in a much shorter time and so do not manifest themselves in the same way under normal experimental conditions. In these cases a low frequency, ~ 20–50 Hz, applied voltage and a fairly simple a.c. bridge serve admirably to measure the resistance, without any complications arising from electrode polarization.

3.3. *Effect of composition on resistivity*
The electrical resistivity is one of the few physical parameters of silicate glasses that changes markedly with the composition. Glass technologists can manipulate compositions to produce glasses with a resistivity anywhere in the range $\sim 10^{17}$ to $\sim 10^{5}$ Ω m at room temperature, but this is still done largely on the basis of experience or by trial and error. Although there are relatively few detailed systematic studies of the change in resistivity with composition, some general trends are well established. The mobile ions are almost invariably the alkali metal cations. In the extreme case of ' pure ' silica the finite conductivity is due to the presence of small concentrations of alkali metal ions as impurities ($\gtrsim 0.1$ per cent). The resistivity of a glass depends on

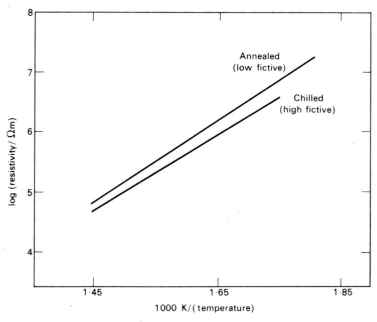

Fig. 3.4. Effect of fictive temperature on resistivity : results for chilled and annealed samples of the same (lead–potash–silica) glass.

54

the number of mobile ions present and also on the ease with which they can jump from one hole in the network to the next. Thus for similar low concentrations, Li^+ or Na^+ ions result in much lower resistivities than the larger K^+ ion (fig. 3.3 a). The addition of divalent modifier ions such as Ca^{2+}, Ba^{2+}, Pb^{2+}, etc. leads to a sharp increase in the resistivity (fig. 3.3 b) ; apparently the large divalent ions block some of the holes in the network, making it more difficult for the alkali ions to migrate. On the other hand for a given glass the resistivity is lower for samples which have a higher fictive temperature ; samples which have been cooled rapidly through the transformation range have a more open network structure (higher fictive temperature) and this makes it easier for ions to migrate (fig. 3.4).

The few investigations that have been carried out on the effect of varying systematically the composition of simple glasses suggest that the quantitative relationship between the concentration of alkali ions

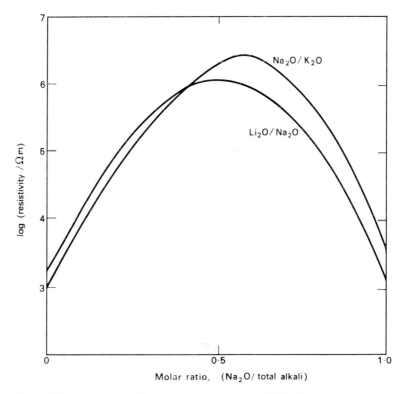

Fig. 3.5. The ' mixed-alkali effect ' ; resistivity at 300°C for a series of glasses containing the same molecular proportion of total alkali, $(Na_2O + K_2O)$ or $(Li_2O + Na_2O)$.

and the resistivity is far from simple. Even in binary soda–silica glasses not all the sodium ions are mobile and there is no obvious relationship between the number which are mobile, and so contribute to the current, and the total number present. When two different alkali metal ions are present in the same glass, the situation is even more involved. The resistivity varies quite markedly with the *ratio* of the concentrations of the two kinds of alkali metal ion even though the total concentration is held constant. Figure 3.5 illustrates the change in resistivity at 300°C which occurs as the molar proportion of Li_2O/Na_2O and K_2O/Na_2O is changed in a series of glasses all containing a total of 33(mol)% alkali metal oxide, i.e. $R_2O . 2SiO_2$, where $R_2O = (Li_2O + Na_2O)$ or $(K_2O + Na_2O)$. Clearly the different species of alkali metal ions present in a complex glass do not contribute independently to the

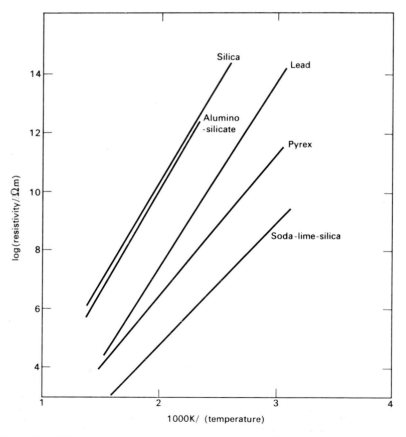

Fig. 3.6. Effect of temperature on the resistivity of some commercial glasses.

56

conductivity. Indeed for simple soda-glasses the resistivity is increased *more* by replacing half the Na_2O by K_2O *or* Li_2O than by simply removing half the Na_2O. A number of other physical properties, especially the viscosity and the internal friction (see Chapter 5), also exhibit large changes with the ratio of alkali concentration in ' mixed-alkali ' glasses. As yet, however, there is no satisfactory quantitative explanation of these effects.

3.4. *Effect of temperature on resistivity*

The resistivity of glasses varies rapidly with temperature. In fact for most glasses

$$\rho = A \exp (E/kT) \qquad (3.1)$$

or

$$\ln \rho = A' + E/kT \qquad (3.2)$$

where A and E are constants characteristic of the particular glass, k is the Boltzmann constant and T is the absolute temperature. Figure 3.6 shows the variation of $\ln \rho$ with $1/T$ for several commercial glasses. Typically for silica-based glasses E lies between ~ 0.5 and 1.0 eV. (kT at room temperature is ~ 0.025 eV $\simeq 10^{-21}$ joule) and A between $\sim 10^{-1}$ and 10^{-4} Ω m. In a general way E is related to the ease with which the mobile ions can jump from one hole to the next, while A depends on the number of mobile ions, the average distance from one hole to the next (i.e. the jump distance) and the number of holes available.

In the following section we shall show how equation (3.1) arises, but as in the quantitative discussion of thermal properties, we shall discuss a model appropriate to a crystalline material and then later contrast this with the situation in a glass. It may be useful first to digress briefly in order to sketch the background to the processes which enable an electric current to flow in an ionic crystal.

3.5. *Ionic conductivity in crystals*

Conduction mechanism in ionic crystals

Ionic crystals (e.g. NaCl, AgBr, MgO, etc.) are also electrolytic conductors ; they too obey equation (3.1). Ion migration can occur in crystals because there are always imperfections (*lattice defects*) in the structure. In the silver halides, for example, some silver ions can be displaced from their normal positions in the crystal, leaving vacant sites, *vacancies*, behind and occupying some of the small spaces, *interstitial sites*, between the other ions in the cubic array (fig. 3.7). Progressive movement of the displaced, *interstitial*, silver ions from one interstitial site to the next under the influence of an applied electric field gives the electric current.

In some crystals vacancies exist by themselves without an equivalent number of interstitials and we can regard the missing atoms or ions as having been moved to normal lattice sites on the surface of the crystal. Vacancies can move around in a crystal. A vacancy in an ionic crystal is displaced from its initial site when one of the ions of the same polarity as the missing one moves from an adjacent site into the vacant one. Vacancy migration in an ionic crystal therefore involves the movement of ions so that this process also can provide a mechanism for the conduction of an electric current.

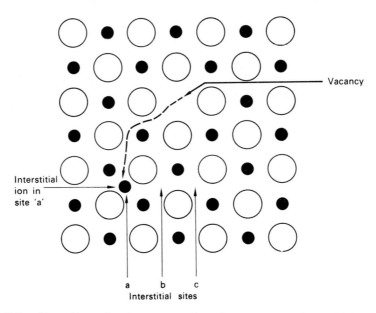

Fig. 3.7. Two-dimensional representation of a vacancy, an interstitial ion and interstitial sites.

The energy of a crystal is increased when vacancies or interstitials are present. Nevertheless at high temperatures a number of vacancies or vacancy–interstitial pairs will exist in thermodynamic equilibrium in a crystal. The equilibrium concentration increases with temperature. If the temperature is high enough for migration to occur, further increase in temperature will result in more vacancies being created at the surface and migrating into the crystal, or, depending on the nature of the crystal, in the creation of more vacancy–interstitial pairs. In some crystals the extra energy required to create these defects causes a detectable increase in specific heat capacity at high temperatures ; the heat capacity is above $3N_A k$ per mole in this case, because vibrations

58

of the atoms are not the only absorbers of energy. When the temperature of a crystal is reduced, the concentration of these defects falls. Vacancies and interstitials recombine or vacancies migrate to the crystal surface. Eventually migration becomes very slow or ceases and then further reduction in the temperature will not produce a corresponding reduction in the concentration of defects. The concentration of vacancies or vacancy–interstitial pairs in ionic crystals at room temperature is *not* the equilibrium one but depends on how rapidly the crystal was cooled from high temperature.

Although in principle both the cations and the anions in an ionic crystal may form interstitials and vacancies, in some crystals the negatively charged anions are so much larger than the cations that the energy required to squeeze the anion into an interstitial site is prohibitive.

Quantitative model of the conduction process in an ionic crystal

We can construct a simple one-dimensional model that will illustrate the essential physics of the conduction process due to the movement of interstitial ions. This will serve as a framework both here and in later chapters to discuss some of those features of the behaviour of glasses which involve internal ion migration.

We shall suppose that the potential energy of one of the interstitial silver ions in a silver halide crystal varies with its position along one direction in the crystal as shown in fig. 3.8 *a*. The ion has a minimum

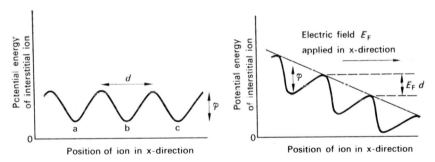

Fig. 3.8. (*a*) Variation of the potential energy of an interstitial ion in a crystal, and (*b*) the influence of an applied electric field.

energy in positions, a, b, c, etc. so that these will correspond to the interstitial sites (like a, b, c in fig. 3.7). For a simple crystalline material the spacing between these positions and also the amplitude of the potential energy variation will be constant. Suppose the interstitial ion is in site b. Its thermal energy will cause it to vibrate about the minimum in the potential well in the same way as for ions on normal lattice sites, although the value of the force constant β will not be exactly the same and the vibration frequency of the interstitial ion and its immediate neighbours may differ slightly from that of ions in normal sites. The *average* vibrational energy of the interstitial ion is fixed by the temperature (for $T \gg h\nu/k$, the average energy will be $3kT$), but the instantaneous value may fluctuate about this average. These fluctuations about the average occur at random but large deviations from the average value occur much less frequently than small ones. The equation giving the probability that the vibrating ion has an energy between E and $E + \delta E$ associated with it is

$$\text{probability} = B \exp{(-E/kT)}\delta E \qquad (3.3)$$

and is called the Boltzmann equation ; k is the Boltzmann constant ($=1\cdot38 \times 10^{-23}$ J K^{-1}) and T is the temperature in kelvin. This equation is widely used in physics. It is the basic equation which leads to the Maxwell–Boltzmann equation for the distribution of the molecular speeds in a gas ; the fluctuation in the speed, and therefore energy, of a given molecule in a gas about the average value is exactly analogous to the fluctuations in the vibrational thermal energy of the ions in our crystal.

The factor B in the Boltzmann equation is a function of temperature and of the physical details of the system being considered. In our particular example it is a certainty (the probability $= 1$) that the vibrating ion has some value of energy between zero and infinity, so that the value of B is determined by the condition that the integral of the expression on the right-hand side of equation (3.3) must be unity.

For an ion vibrating in a potential well, like that around site b in fig. 3.8 *a*, there is a probability that an occasional fluctuation in energy will be large enough for the ion to surmount the barrier and jump into an adjacent well. From equation (3.3) we can deduce that for a fraction of the time $\approx \exp{(-\phi/kT)}$ the ion will have an energy equal to or greater than ϕ. In the sort of situation we are concerned with, where $\phi \gg kT$, this is a very small fraction : if $T \sim 700$ K and $\phi \sim 1$ eV, it is $\sim e^{-20} \approx 10^{-9}$. The ion ' collides ' with each barrier on opposite sides of the potential well once in every vibration and statistically for a fraction of these ' collisions ' the energy will be high enough for the ion to pass over the barrier. This fraction of ' collisions ' is the same as the fraction of the *time*, i.e. $\approx \exp{(-\phi/kT)}$, so that the statistical number of jumps per second is simply

$$\nu \exp{(-\phi/kT)}, \qquad (3.4)$$

where ν is the frequency of vibration of the ion. The height of the potential barrier, ϕ, is called the *activation energy*. For an interstitial ion, ν will be of the same order as the Einstein frequency for normal ions in the lattice, $\sim 10^{13}$ Hz.

We restrict our model for the conduction process to one dimension so that the ions vibrate and jump in the x-direction only. Let us suppose that we have a long chain of such potential wells and many interstitial ions but that the number of interstitials is much less than the number of wells so that there is no mutual interaction between the interstitial ions. Ion jumping will be frequent when ϕ/kT is small, but there will be no *net* displacement of charge since as many ions will jump to the right as to the left.

When we apply an electric field in the x-direction, the potential energy of a positively charged interstitial ion changes to that shown in fig. 3.8 b, because we must superpose on the previous distribution the energy of the ion in the externally applied field. Now the heights of the potential barriers to the left and to the right of the ion are no longer the same ; they are respectively $(\phi + \frac{1}{2}E_{F}qd)$ and $(\phi - \frac{1}{2}E_{F}qd)$. E_{F} is the applied field strength, d the separation between the interstitial sites, and q is the charge on the ion. A given ion is now more likely to jump to the right than to the left since

$$\nu \exp\,[(-\phi + \tfrac{1}{2}E_{F}qd)/kT] > \nu \exp\,[(-\phi - \tfrac{1}{2}E_{F}qd)/kT].$$

The net number of ions passing a given point per second in the direction of the applied field is therefore

$$\sim n\nu \exp\,(-\phi/kT)[\exp\,(\tfrac{1}{2}E_{F}qd/kT) - \exp\,(-\tfrac{1}{2}E_{F}qd/kT)],$$

where n is the number of ions per well ($\ll 1$). Since $E_{F}qd/2kT$ is normally very small and $e^{x} \approx 1 + x$ if $x \ll 1$, this can be written

$$\sim n\nu(E_{F}qd/kT)\exp\,(-\phi/kT).$$

The electric current in the direction of the applied field is just the net number of charges passing a given point in one second multiplied by the charge. Therefore the current is given by

$$I = \frac{q^{2}n\nu E_{F}d}{kT}\exp\,(-\phi/kT). \qquad (3.5)$$

We may note that $I \propto E_{F}$ at constant T, i.e. our model ' obeys ' Ohm's law. And since the resistivity

$$\rho = \frac{Ra^{2}}{l} = \frac{V}{I}\frac{a^{2}}{l} = \frac{E_{F}a^{2}}{I}$$

where a^{2} is the effective cross-sectional area and l is the length of our one-dimensional chain,

$$\rho \approx \frac{kTa^{2}}{q^{2}n\nu d}\exp\,(\phi/kT). \qquad (3.6)$$

In extending this calculation to a three-dimensional model, we should have to allow for changes in barrier height and for jumping of the ions

61

in directions other than exactly parallel to the applied field. In fact this does not alter the expression for ρ beyond the introduction of a small numerical constant, so that for a similar three-dimensional model we get

$$\rho \approx \frac{CkT}{q^2 n_v v d^2} \exp{(\phi/kT)}, \tag{3.7}$$

where C is a numerical constant (~ 4) and n_v ($\simeq n/a^2 d$) is the number of interstitial ions per unit volume.

For a pure crystal, the number of interstitial ions in equilibrium at any given temperature is also determined by the Boltzmann equation so that

$$n_v = D \exp{(-\gamma/kT)},$$

where γ is simply related to the energy required to form the interstitial ion. Then

$$\rho \approx \frac{CkT}{q^2 D v d^2} \exp{[(\phi+\gamma)/kT]} \tag{3.8}$$

or

$$\rho \approx A_0 T \exp{[(\phi+\gamma)/kT]}, \quad A_0 = \text{constant}.$$

Comparison with experiment

Our model yields a relationship which is slightly different from that found by experiment, compare equation (3.1), for ionic crystals. The model predicts that the pre-exponential term depends on the temperature. Evidently we should expect $(\ln \rho - \ln T) = \ln A_0 + (\phi+\gamma)/kT$. In practice however experiments can only be carried out over a limited temperature range, high enough to produce a measurable ρ and yet below the melting point. If T ranges from ~ 400 K to 700 K, typically $\ln \rho$ changes from ~ 24 to 12 while $\ln T$ changes from 6 to 6·6. Thus over a practicable range of temperature $\ln T$ is essentially constant and we can put $(\ln A_0 + \ln T) = \ln A = \text{(constant)}$. We have then

$$\rho = A \exp{[(\phi+\gamma)/kT]}, \tag{3.9}$$

where A is essentially constant and depends on (equation 3.8) the charge on the mobile ion, the vibration frequency and the jump distance; ϕ is the activation energy for jumping and γ the activation energy for formation of an interstitial ion in the crystal. The total activation energy for ionic conduction is therefore (jump energy) + (effective formation energy).

In some other types of ionic crystal, electrical conduction takes place by the drifting of vacant cation sites. A model for this conduction process can be constructed which exactly parallels the one above; ϕ and γ would become the activation energies for vacancy jumping and formation respectively.

Determination of ϕ and γ

It is possible to create a high concentration of interstitials or vacancies in many ionic crystals by adding impurities. For example, if we grow crystals from molten KCl to which a small amount of $CaCl_2$ has been added, the structure of the crystals is much the same as for pure KCl, in which there are an equal number of positive ion and negative ion sites. The divalent impurity ions (Ca^{2+}) occupy a few of the positive ion sites, which normally would be occupied by K^+ ions, and an equal number of normal K^+ ion sites are left vacant. The concentration of these vacancies is determined by the amount of $CaCl_2$ and is independent of the temperature. At low temperatures these crystals show much higher conductivities (lower resistivities) because more vacancies are available ; the resistivity is given by

$$\rho = A' \exp (\phi/kT).$$

The term in γ/kT is missing because the number of vacancies is controlled by the impurity concentration rather than by the Boltzmann

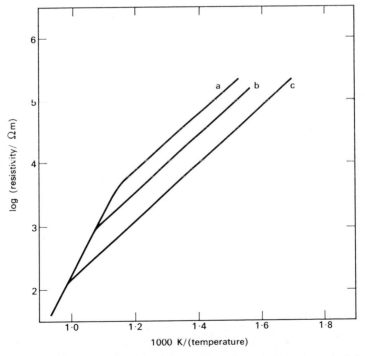

Fig. 3.9. Effect of temperature on the resistivity of KCl crystals ' doped ' with traces of $BaCl_2$: the mole fraction of $BaCl_2$ is approximately $1\cdot3 \times 10^{-5}$, 3×10^{-5} and 9×10^{-5} for a, b and c respectively.

factor. At high temperatures where γ/kT becomes small, additional vacancies will be created by thermal excitation. The variation of $\ln \rho$ with $1/T$ will therefore show two distinct regions (fig. 3.9). At low temperatures the vacancy concentration is determined by the number of divalent impurity ions ; γ/kT is so large that changes in temperature do not significantly change the vacancy concentration and the slope of the line $\ln \rho$ versus $1/T$ is ϕ/k. On the other hand, at high temperatures, γ/kT is smaller and the vacancy concentration changes with temperature and the slope becomes $(\phi+\gamma)/k$. Using tricks of this kind it is possible to determine the values of ϕ and γ separately from measurements of resistivity as a function of temperature.

A most important point as far as we are concerned here is that the activation energies, determined by plotting $\ln \rho$ versus $1/T$ from experimentally measured data on crystals, are related directly to the energies for jumping and formation of the individual charge carriers. Provided the concentration of interstitial ions (or vacancies) is not too high, there is no mutual interaction between them and *in a crystal* each interstitial ion or vacancy is in an identical environment : each ion has the same barrier to overcome in order to jump into the adjacent equivalent site. The same is not true for the sodium ions in a glass !

3.6. *Comparison of conductivity in crystals and glasses*

The electrical conductivity of glasses is due to the migration of alkali metal ions. However there is, as yet, no general agreement on the detailed *mechanism* involved. It is not known, for example, whether the migration mechanism could be considered analogous to the movement of interstitial ions or to the movement of vacancies in ionic crystals. Are some of the Na^+ or other alkali metal ions displaced into ' higher energy sites ' and do they then jump from one of these into another ? Or do the ions simply jump from one ' normal ' site into an adjacent one ? If the latter process is possible, what determines the number of vacant ' normal ' sites ? Is the number of vacant sites dependent on the temperature (and given by the Boltzmann equation) or is it controlled by composition ?

Without the answers to these questions we do not know whether the activation energy E found experimentally is the sum of a jumping energy and a formation energy or just a jumping energy. There is another serious, fundamental difficulty. The one-dimensional model of fig. 3.8 used to discuss the conductivity of ionic crystals is inappropriate for glasses. A more realistic model in keeping with the random network theory for the structure of glasses is shown in fig. 3.10. Neither the spacing of the wells nor the barrier height will be constant, irrespective of whether the wells correspond to ' normal ' or to ' interstitial ' sites. Passage of a steady electric current involves the migration of ions right through the network so that the contribution to E

from the 'jumping energy' will be a complex average of the individual barrier heights.

Changes in composition have a marked influence on both E and A in equation (3.1), but in general it is not possible to offer an interpretation of these changes. Even in the simple binary glasses, soda–silica and potash–silica, E changes not only with the nature of the mobile ion but also with the number of alkali metal ions present, fig. 3.11. Evidently, increasing the number of Na^+ or K^+ ions in the glass does much more than provide extra charge carriers; the fall in E means that it is much easier for the ions to move through the network. In this case it is plausible that the fall in E with alkali content is due to the 'loosening' of the network as more and more SiO_4 tetrahedra have first one and then two corners detached. It is easier to push an alkali

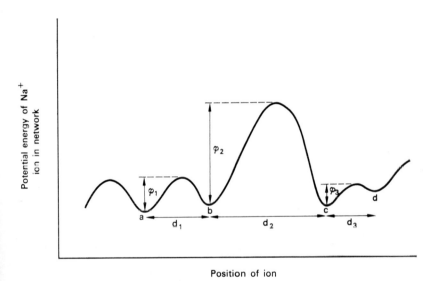

Fig. 3.10. Potential energy of a mobile ion in a glass.

metal ion through a 'window' into an adjacent hole if the tetrahedra which form the 'window frame' are not rigidly fastened by all four corners to the rest of the network. This simple argument is not one of those grossly oversimplified pictures sometimes used in an introductory account of a sophisticated theory; it reflects not unfairly the present 'state of the art'.

Compared with the detailed interpretations possible for ionic crystals, those for glasses are extremely vague and naive. It is not at all difficult to expose current ignorance. We have only to introduce an apparently

65

quite minor change in composition in simple alkali–silica glasses to produce dramatic effects which are still inexplicable.

For example the glass of composition Na_2O, $2SiO_2$ has $A \sim 10^{-3}$ Ω m and $E \sim 0.65$ eV. If we replace the Na_2O by K_2O, A and E are not much changed but replace *half* the soda by potash, and A decreases by $\sim 10^3$ and E is nearly doubled ; we still do not know why.

Dielectric behaviour

3.7. *Introduction*

When an insulating material is placed in an electric field, it becomes polarized : the positive and negative charges within the material are displaced relative to one another into new equilibrium positions. Any

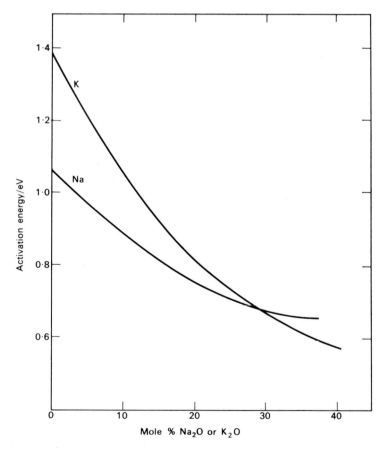

Fig. 3.11. Effect of composition on the activation energy for ionic conduction in the simple binary glasses $SiO_2 + Na_2O$ and $SiO_2 + K_2O$.

volume element within a homogeneous isotropic material acquires an electric dipole moment and surface charges appear at the boundaries of the material where the outermost layer is depleted of either positive or negative charges (fig. 3.12).

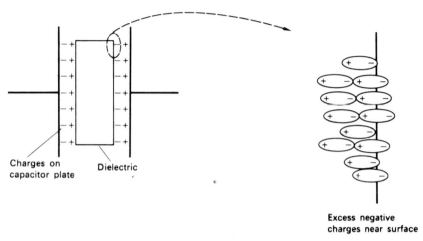

Fig. 3.12. Polarization of a dielectric.

Three distinct physical processes can lead to relative displacement of the charges within a material in an applied electric field.

(1) The electron clouds around individual nuclei can be displaced so that the nucleus is no longer in the exact centre and each atom or ion acquires an induced electric dipole moment.

(2) Permanent electric dipoles which already exist in the material (usually due to the presence of discrete molecules with an asymmetric distribution of charge) may be rotated by the electric field from normally random orientations into directions more nearly parallel to the field.

(3) Positive and negative ions existing as such within the material may undergo relative displacement in the electric field.

The three processes are illustrated schematically in fig. 3.13. The first process occurs in all materials but obviously the second and third can occur only if the nature of the material is such that permanent dipoles or ions, respectively, are present. Even when permanent dipoles are present, their rotation may be inhibited and this is often the case in solids. The dipole moment per unit volume is called the *polarization* and it is, usually, proportional to the strength of the applied electric field. The constant of proportionality, the dielectric susceptibility,

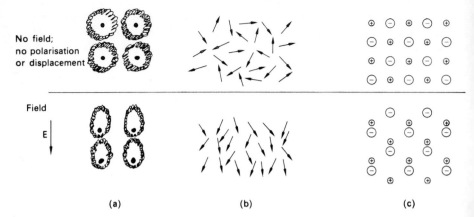

Fig. 3.13. To illustrate mechanisms of charge displacement and polarization of a dielectric : (*a*) distortion of electron clouds, (*b*) reorientation of permanent dipoles, (*c*) displacement of ions.

varies from one material to another and is much larger for those materials which exhibit polarization by dipole orientation and ionic displacement.

3.8. *Dielectric properties*

In a homogeneous isotropic medium the polarization, P, is parallel to the applied electric field, E, and

$$P = \epsilon_0 \chi E = \epsilon_0(\epsilon_r - 1)E, \qquad (3.10)$$

where χ is the dielectric susceptibility, ϵ_r the *relative permittivity* (dielectric constant) of the medium, and ϵ_0 is the permittivity of a vacuum. The value of ϵ_r is commonly used to provide a measure of the polarizability of a material ; its value is an important characteristic of any dielectric. There are two other ' dielectric properties ' which are also of interest and are important in practice. In most dielectric materials the charge displacements which produce the polarization do not all take place at the same rate. The distortion of the electron clouds around individual nuclei occurs in $\sim 10^{-15}$ s, the displacement of ions *within* their potential wells takes $\sim 10^{-13}$ s but ionic displacement which involves jumping over the barrier into an adjacent site, or the reorientation of permanent dipoles, is very much slower and may take from 10^{-6} s to minutes or even hours depending on the particular process, the material and the temperature. As a steady electric field is applied to a dielectric, a polarization occurs almost instantaneously, but this may be followed by a further slow increase due to the rearrangement of charges by the slower processes. This *delayed polarization* gives rise to a number of effects which, because of the widely different time scales involved, often appear to be quite separate phenomena. The 'transient conductivity' referred to in the earlier part of this chapter can be regarded as a delayed polarization with a very long time constant.

In an alternating field of high frequency, the slow processes may not contribute to the polarization at all because the applied field reverses direction before they occur. The permittivity of a material may fall off therefore as the frequency increases. At intermediate frequencies the slow processes cause the total polarization to lag behind the applied field ; maximum polarization will not occur at the same points in time as the maxima in the field. The energy expended in polarizing the dielectric is not all recovered when the applied field returns to zero. If the total polarization at any instant is plotted against the field at the same instant, a hysteresis loop is obtained and the area of the loop is related to the energy lost per cycle. Some energy will also be dissipated because the ordinary 'd.c. conductivity' of real materials is not zero and a small current will flow in the applied field and will produce Joule (V^2/R) heating. The magnitude of these energy losses (known as dielectric losses) is an important property of any dielectric material. This property is represented by a parameter called the *dissipation factor*, which is written as tan δ, and the power loss per unit volume of a dielectric is proportional to ϵ_r tan δ.

The third important characteristic of a dielectric is the electric field strength at which the high resistance (i.e. the insulating character) of the material is destroyed. The field at which insulation breakdown occurs is called the *dielectric strength*.

All three of these dielectric properties, the relative permittivity, the dissipation factor and the dielectric strength, may change with frequency. Some experimental values of these parameters are given in table 3.1.

3.8. *Capacitors*

For a parallel-plate capacitor in a vacuum the capacitance is given by

$$\frac{Q}{V} = C_0 = \epsilon_0 A/d, \qquad (3.11)$$

where A is the area of each plate, d is the distance between the two plates and ϵ_0 is the permittivity of a vacuum, $8 \cdot 85 \times 10^{-12}$ F m^{-1}. When the space between the plates is filled with an insulating material (a dielectric) then for a given charge, Q, on the plates, the potential difference, V, is reduced by the surface charges induced at the boundaries of the dielectric. The capacitance becomes

$$C = \epsilon_0 \epsilon_r A/d = \epsilon_r C_0, \qquad (3.12)$$

where ϵ_r is the relative permittivity of the material between the plates.

F

Material	Dielectric relative permittivity ϵ_r	Dissipation factor $\tan \delta \times 10^4$	Frequency	Dielectric strength/V m^{-1}
Mica	7	2	1 MHz	$\sim 10^8$–10^9
Perspex	2·6	160	1 MHz	$\sim 4 \times 10^8$
	3·4	600	10^2 Hz	
Water	80	650	10^3 MHz	
Paper (capacitor tissue)	2·3	22	10^3 Hz	$\sim 10^7$
Porcelain	7	70	1 MHz	6×10^6
Silica glass	3·8	2	10^3–10^6 Hz	7×10^8
Borosilicate glass (Pyrex)	4·8	128	10^2 Hz	
	4·6	46	1 MHz	
Lead–silicate glass	6·7	78	10^2 Hz	
	6·4	16	1 MHz	
Soda–lime–silica glass	8·3	780	10^2 Hz	
	6·9	100	1 MHz	
NaCl	5·62	—	Static	
	2·75	—	Optical	
NaF	6·0	—	Static	
	1·74	—	Optical	
LiCl	11·05	—	Static	
	2·75	—	Optical	

Table 3.1. Dielectric properties of glasses and other materials.

When a sinusoidally varying potential is applied to a capacitor the current which flows in the circuit also varies sinusoidally. The charge on the capacitor is given by

$$Q = CV \qquad (3.13)$$

and for an *ideal* capacitor the capacitance, C, is constant, so that

$$I = dQ/dt = C \, dV/dt \; ; \qquad (3.14)$$

if $V = V_0 \sin \omega t$, then

$$I = V_0 \omega C \cos \omega t = V_0 \omega C \sin (\omega t + \pi/2). \qquad (3.15)$$

Thus the current leads the voltage by a phase angle of $\pi/2$ (a quarter of a period), fig. 3.14.

With a real dielectric between the plates of the capacitor, there will be a component of the total current which is *in phase* with the applied voltage due to the small ' leakage ' currents and to the delayed polarization in the dielectric. The total current is not therefore exactly $\pi/2$

70

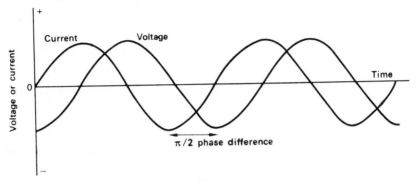

Fig. 3.14. Variations with time of voltage and current for an ideal capacitor.

ahead of the voltage. We write the actual phase angle as $(\pi/2 - \delta)$, then for

$$V = V_0 \sin \omega t,$$

we have

$$I = I_0 \sin [\omega t + (\pi/2 - \delta)] = I_0 \cos (\omega t - \delta). \qquad (3.16)$$

This can be written

$$I = (I_0 \cos \delta) \cos \omega t + (I_0 \sin \delta) \sin \omega t. \qquad (3.17)$$

The first term is the component which is $\pi/2$ out of phase with the applied voltage and the second term is the component which is in phase with the voltage. As in the ideal case, the capacitor plates are charged and discharged by that part of the total current which is out of phase with the voltage. The amplitude of this component is determined by the capacitance, i.e.

$$I_0 \cos \delta = V_0 \omega C. \qquad (3.18)$$

The component of current which is in phase with the voltage will dissipate energy at a rate given by the usual relationship, VI. The average power lost is therefore

$$V_0 I_0 \overline{\sin \delta \sin^2 \omega t} = \tfrac{1}{2} V_0 I_0 \sin \delta \qquad (3.19)$$
$$= V' I' \sin \delta,$$

where V' and I' are the root-mean-square values of voltage and current. The phase difference angle, δ, is known as the *loss angle* of the capacitor and $\sin \delta$ is the *power factor*. The loss angle is determined by the behaviour of the dielectric and the electrical power lost in the capacitor is dissipated in the dielectric and ends up as heat.

71

From equations (3.18) and (3.19) the power lost is

$$\tfrac{1}{2}V_0{}^2\omega C \tan \delta,$$

but $C = \epsilon_0\epsilon_r A/d$ and the volume of the dielectric is Ad ; therefore the energy dissipated per unit time per unit volume of dielectric is

$$\tfrac{1}{2}(V_0{}^2/d^2)\epsilon_0\epsilon_r\omega \tan \delta$$

or

$$\tfrac{1}{2}E_0\epsilon_0\epsilon_r\omega \tan \delta, \qquad (3.20)$$

where E_0 is the amplitude of the alternating electric field. The loss angle is a fundamental property of the dielectric. But, because of the form of these expressions for the energy losses, it is more usual and convenient in practice to use the value of $\tan \delta$ (the dissipation factor) to characterize a material. In fact for useful dielectrics δ is very small and $\tan \delta \approx \sin \delta \approx \delta$.

In an a.c. circuit, a real capacitor is equivalent to an ideal capacitor with a large resistance in parallel with it. The power loss in the real capacitor is equivalent to Joule heating in the shunting resistance, R_s. Thus

$$\tfrac{1}{2}V_0{}^2/R_s = \tfrac{1}{2}V_0{}^2\omega C \tan \delta$$

and therefore

$$R_s = 1/\omega C \tan \delta. \qquad (3.21)$$

3.10. *Experimental methods*

The permittivity and dissipation factor of dielectrics are usually determined, over a wide frequency range from d.c. to radio frequencies,

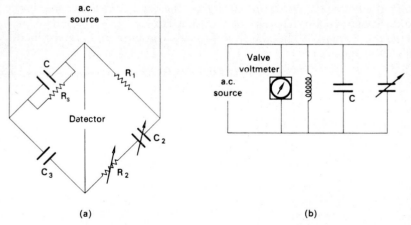

(a)

(b)

Fig. 3.15. Circuits for measurement of an unknown capacitance, C. (a) Schering a.c. bridge ; (b) resonance method.

72

by measuring the capacitance and effective shunting resistance of a capacitor filled with the material. A capacitor consisting of a pair of metal disc electrodes is normally used for solid dielectrics. One of the discs is surrounded by a guard-ring to reduce edge effects and a thin sample of the dielectric material is placed on this and clamped by the other electrode disc.

At audio frequencies the capacitance and equivalent shunting resistance are measured using an a.c. bridge. The balancing of an a.c. bridge requires that two conditions be satisfied simultaneously ; the amplitudes and phases of the voltages at the opposite corners of the bridge must be the same in order to minimize the detector current. The Schering a.c. bridge suitable for the measurement of capacitance is shown in fig. 3.15 *a*. The unknown lossy capacitor is represented by the combination C and R_s. C_3 is a standard capacitor of known value similar in magnitude to C, and C_2 is usually a calibrated, variable, air capacitor of negligible leakage resistance. The balance conditions for this bridge can be shown to be

$$R_s = R_1(C_2/C_3)$$

and

$$C = C_3(R_2/R_1).$$

These conditions are independent of one another and the capacitance C is determined in terms of the standard C_3 and the ratio of resistances; tan δ is given by $(\omega C R_s)^{-1}$.

The capacitance of the cell containing the dielectric is normally very small since it is inconvenient to make the area of the plate larger than a few square centimetres. In the measurement of very small capacitances the existence of stray capacitances between different parts of the circuit or between the circuit and earth becomes troublesome, and a special procedure is used to eliminate the effects due to these unknown strays. At radio frequencies, where a.c. bridges are very awkward to use for accurate measurements, the capacitance of the cell containing the disc of dielectric may be determined by incorporating the capacitor in a resonance circuit (fig. 3.15 *b*). If the circuit containing the dielectric cell in parallel with a standard variable capacitor is tuned to resonance and then the dielectric is removed, the change in capacitance can be ascertained from the adjustment of the variable capacitor required to restore resonance. The energy dissipated by the dielectric can be found from the change in the voltage amplitude at resonance.

The intrinsic dielectric strength of glasses is probably of the order of 10^9 V m^{-1} but in practice such strengths are seldom observed. The experimental measurement of the field strength for intrinsic breakdown of such high strength materials is particularly difficult and we shall not discuss any of the details here. Elaborate precautions must be taken in order to avoid distortion of the electric field near the

73

edges of electrodes.   Since the breakdown of air at normal pressures occurs at about 3×10^6 V m^{-1}, high strength dielectrics are often immersed in oil ; even then breakdown or leakage currents in the oil can lead to local heating which initiates breakdown at the edges of the dielectric material.

3.11. Theories of dielectric behaviour

Effective field and the Clausius–Mossotti Equation

In principle the polarization induced in a material by an applied electric field can be calculated from its structure and the polarizability of its constituent molecules, atoms or ions.   In practice however there are major difficulties which, for most of the technically important dielectrics, have not been overcome.   One of the complications is that the field which polarizes an atom *inside* the material is not, in general, the same as the externally applied field.   The field, E_I, acting on an atom inside a material, is the resultant of the externally applied field, E_A, and the fields produced by the surrounding dipoles.   For non-polar liquids and gases and for simple atomic or molecular crystals which have a highly symmetrical structure, a useful approximation can be made in order to calculate the internal field.   It turns out that

$$E_I = E_A + P/3\epsilon_0. \qquad (3.22)$$

For a material consisting of atoms or molecules all of the same kind we have

$$P = n\alpha E_I, \qquad (3.23)$$

where n is the number of atoms or molecules per unit volume and α is their polarizability.   Combining equations (3.10), (3.22) and (3.23) leads to an equation relating the macroscopically observable permittivity of the material and the molecular polarizability :

$$\frac{\epsilon_r - 1}{\epsilon_r + 2} = \frac{n\alpha}{3\epsilon_0} = \frac{N_A \rho}{M} \frac{\alpha}{3\epsilon_0} \qquad (3.24)$$

where N_A is the Avogadro constant
      M is the mass of one mole
and ρ is the density.
This equation is known as the Clausius–Mossotti equation.   It can be extended to materials containing an intimate mixture of more than one kind of atom or molecule by replacing $n\alpha$ by $n_1\alpha_1 + n_2\alpha_2 + \ldots$, where n_1, α_1, etc. refer to the different constituents of the ' mixture '.

Phenomenological description of dielectric behaviour

The situation is considerably more complicated for an ionic solid, since here the relative displacement of the ions also contributes to the polarization.   Ionic displacement can produce a larger effect than the

74

polarization of the individual ions but the magnitude of the contribution depends on the relative sizes of the ions present (*cf.* values for LiF and NaF in table 3.1) and on the details of the structure of the solid.

For glasses, and indeed for many other technically important dielectrics, the irregularity of structure and the variety of atoms or ions present make it impossible to calculate the macroscopic polarization and thus the relative permittivity from the properties of the ions. The dielectric behaviour of these more complex materials can be *described* quantitatively by means of a phenomenological theory. It often happens that, even though we do not know in detail the nature or the scale of the individual atomic or sub-atomic events responsible for a series of experimentally observed effects, it can be shown that the effects are related and that they can all be described quantitatively by equations derived from a single mathematical model. The equations contain one or more parameters which are characteristic of a particular material and can be determined by experiment, but which cannot necessarily be related to the fundamental properties of matter.

This sort of approach is used frequently in Physics ; it provides a convenient way of summarizing experimental observations in a mathematical form ; it facilitates the comparison of the behaviour of different materials and may even lead to the prediction of new effects. A phenomenological theory has often pointed the way to a more fundamental understanding of a physical phenomenon, but is also useful when, as here, the fundamental processes responsible for the behaviour are known but the contribution from different types of process in a complex material is difficult or impossible to predict.

The physical processes that contribute to the polarization of a dielectric are essentially of two different kinds. In one kind, the charges are displaced from their original positions against ' elastic ' restoring forces. To a first approximation these forces are proportional to the displacement of the charges. The distortion of the electron clouds around individual atoms and the displacement of ions *within* their potential wells are examples of processes of this kind. In the second kind of process (*relaxation processes*), the distribution of charges or of permanent-dipole orientation is changed by the applied electric field from a random to a more ordered form. Relaxation processes often require an activation energy and they are therefore inherently slower than processes of the first kind. On removal of the external field it is the ' elastic ', electrostatic forces which restore the normal configurations in the first case and it is the disordering effect of thermal agitation (more elegantly, the maximization of the entropy) in the second case.

Relaxation processes

We set up a simple phenomenological theory by assuming that, following the application of an electric field, there is an instantaneous

75

polarization (P_1) proportional to the field, followed by a further slow polarization (P_2) which approaches, exponentially with time, a saturation value P_s. P_s is also assumed to be proportional to the applied field. The assumption of an exponential variation of P_2 with time would be valid for any one of a wide variety of activated processes ; it is equivalent to assuming that the instantaneous rate of change of the delayed polarization (P_2) is proportional to ($P_s - P_2$), i.e.

$$\frac{dP_2}{dt} = \frac{1}{\tau}(P_s - P_2), \tag{3.25}$$

where τ is a constant, so that $P_2 = P_s[1 - \exp(-t/\tau)]$ for E and therefore P_s constant.

The decay of polarization after the removal of an applied field ($E = 0$, $\therefore P_s = 0$) is given by

$$\frac{dP_2}{dt} = \frac{-P_2}{\tau}, \tag{3.26}$$

so that

$$P_2 = P_0 \exp(-t/\tau) \tag{3.27}$$

(see fig. 3.16).

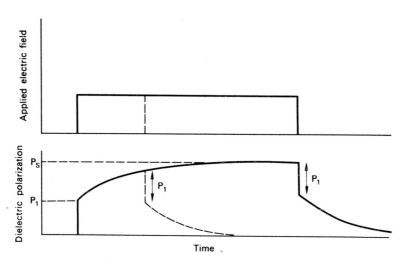

Fig. 3.16.   Instantaneous and delayed polarization.

The constant τ is known as the *relaxation time* and may be used to characterize a particular process giving rise to delayed polarization. Comparison of equations (3.25) and (3.26) with those for radioactive decay shows that τ is simply related to the ' half-life ' for the process.

The time required for the polarization to decay to one-half its initial value *after* removing the field, i.e. $P_1 = 0$ and $P_2 = \frac{1}{2}P_0$, is given by $t = \tau/0 \cdot 693$. The value of τ characterizing a process which involves the excitation of ions over potential energy barriers will vary with temperature. Since τ is inversely proportional to the number of jumps per second, we can expect from equation (3.4) that

$$\tau \propto \exp(\phi/kT). \tag{3.28}$$

In an alternating field of low frequency, such that $\omega = 2\pi\nu \ll 1/\tau$, the relaxation processes occur much faster than the changes in the applied field ; the total polarization will be $(P_1 + P_s)$ and this will keep pace with the changes in the applied field. At high frequencies where $\omega \gg 1/\tau$, the change in P_2 during each half-cycle of applied field is negligible : the polarization will be given by P_1. At intermediate frequencies some delayed polarization will occur but the total polarization will lag behind the applied field. The maximum polarization will be less than $(P_1 + P_s)$ and will not occur at the same point in time as the maximum in the applied field.

Expressions for the relative permittivity and loss factor as a function of frequency can be obtained by solving equation (3.25) with $P_s = E = E_0 \cos \omega t$, i.e. for a periodically varying applied field. The relative permittivity is given by

$$\epsilon_r = \epsilon_\infty + \frac{\epsilon_s - \epsilon_\infty}{(1 + \tau^2\omega^2)} \tag{3.29}$$

and the dissipation factor by

$$\epsilon_r \tan \delta = \frac{(\epsilon_s - \epsilon_\infty)\omega\tau}{(1 + \tau^2\omega^2)}, \tag{3.30}$$

where ϵ_s and ϵ_∞ are the relative permittivities in a static and in a very high frequency applied field, respectively. The variation of permittivity and dissipation factor with frequency according to these expressions is illustrated in fig. 3.17. Significant changes in permittivity and energy losses occur only over a limited range of frequencies, $10/\tau > \omega > 1/10\tau$.

In practice there may be more than one process responsible for delayed polarization in a given material but the theory can be extended so that we can characterize each of the processes by a separate relaxation time.

Effects at very high frequencies
So far we have regarded the charge displacement brought about by the distortion of the electron clouds of the ions and by the displacement of ions within their potential wells as instantaneous. These processes do, however, take a finite time and at very high frequencies we can

expect to see changes in the permittivity and associated energy losses, as these processes become too slow to occur before the field is reversed. Ion displacement is the slower process ; the time required is due to the inertia of the ions rather than to the need to overcome a potential barrier and will be of the order of the natural period of the 'free' vibrations, viz. 10^{-13} s. Alternating electric fields with a frequency approaching *optical* frequencies are required, therefore, before changes

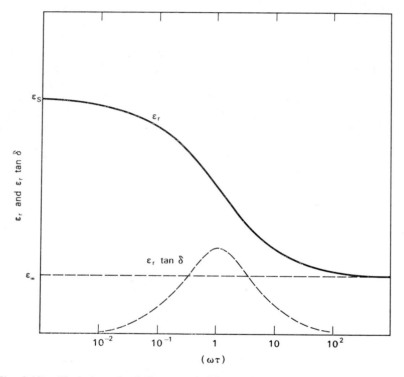

Fig. 3.17. Variation of relative permittivity and dielectric loss with frequency for a dielectric with a single relaxation time.

in permittivity occur due to the loss of the contribution from ionic displacements. Since the alternating electric field associated with an electromagnetic wave will displace the ions and electrons, we can anticipate that these effects will manifest themselves in the *infrared* and *optical* properties of the material. There is an intimate connection between the electrical and optical properties of materials ; the connection is developed formally in Maxwell's electromagnetic theory and we shall discuss it in more detail in the next chapter The theory

78

shows, for example, that the velocity of light in a non-conducting medium is given by

$$c = 1/\sqrt{(\mu_0 \epsilon_0 \mu_r \epsilon_r)},$$

where μ_0 and ϵ_0 are the permeability and the permittivity, respectively, of a vacuum; and μ_r, ϵ_r are the relative permeability and relative permittivity of the medium. Since the refractive index is defined by

$$n = \frac{\text{speed in vacuum}}{\text{speed in medium}}$$

we have

$$n = \sqrt{(\epsilon_r \mu_r)}$$

or for a non-magnetic material ($\mu_r = 1$)

$$n^2 = \epsilon_r. \tag{3.31}$$

This equation holds only for refractive index and relative permittivity determined at the same frequency. We shall not expect the optical refractive index to be given by the square root of the relative permittivity measured at, say, radio frequencies. On the other hand we can infer that changes in refractive index will occur at the same frequencies as changes in permittivity, for example at the resonant frequencies of ions within the material.

3.12. Dielectric properties of glasses

Experimental results

The relative permittivity of silica glasses in static fields ranges from 3·7 for pure silica to ~10 for a high lead glass. The permittivity is approximately proportional to the density, $\epsilon_r/\rho \sim 2\cdot2$, and the high permittivity glasses are those containing high proportions of large, easily polarizable ions such as Ba^{2+} or Pb^{2+} or large numbers of mobile alkali ions. In a general qualitative way these trends correspond to the form of the Clausius–Mossotti equation and we cannot expect more than that for these irregular arrangements of ions.

The dissipation factor (tan δ) is of the order 10^{-4} to 10^{-2} and again is lowest for pure silica and highest for high alkali glasses. At a given frequency and temperature there is in fact a very good correlation between the dissipation factor and the d.c. conductivity of the glass. A ' mixed-alkali ' effect can be used with advantage, as far as dielectric behaviour is concerned, since glasses containing high total alkali concentrations but equal proportions of, for example, Na_2O and K_2O have high values of relative permittivity without correspondingly high dissipation factors.

Most glasses show only a very slow and comparatively small change in permittivity and dissipation factor over a wide range of frequency, d.c.

79

to 10^{10} Hz (fig. 3.18). Raising the temperature increases slightly the permittivity but gives much larger values of tan δ ; for many glasses tan δ is proportional to exp (BT) and the ' constant ' B is larger for lower frequencies.

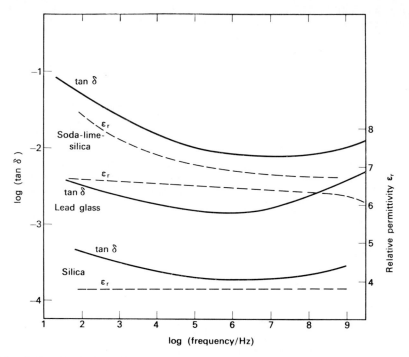

Fig. 3.18.   Variation of relative permittivity, ϵ_r, and dissipation factor, tan δ, for some commercial glasses.

Power losses in practice

The power loss in a dielectric is proportional to the product of ϵ_r and tan δ, so that in selecting glasses for use as insulators for alternating voltages it is important to choose those compositions which have low values of both permittivity and dissipation factor. High chemical durability of the surface may also be important ; a glass insulator which is to be exposed to the atmosphere for a long time must be resistant to attack by water (and atmospheric pollutants) to avoid the attendant increase in surface conductivity, which could lead to breakdown. Glasses from the borosilicate group meet all these requirements and are used as insulators.

On the other hand, for a capacitor dielectric, high permittivity and low dissipation factor are required.   Lead–silicate glasses are suitable for this purpose ; the dielectric properties of 'high-lead' glasses are comparable with those of mica. Very thin ribbons, 2×10^{-2} mm thick, can now be produced and these are used to construct capacitors consisting of alternate sheets of glass and metal foil.

Interpretation of dielectric properties

The variation with frequency of the permittivity and dielectric loss of glasses is more complex than that shown in fig. 3.17. It is possible to represent the observed behaviour using the theory of relaxation processes by assuming that there are a large number of different relaxation times. By far the most important contribution to the dielectric losses in glass arises from the ability of the alkali metal ions to jump from one hole to the next in the network. Thus, basically the same process gives rise to d.c. conductivity, to absorption currents *and* to dielectric loss.

The way this process gives rise to dielectric loss can be illustrated in a simple fashion by reference to fig. 3.19 which shows the effect of

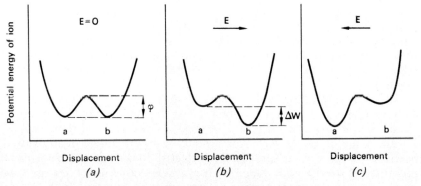

Fig. 3.19.   Effect of an applied electric field on a double potential well.

an applied field on a double potential well. Suppose that in the absence of an applied field, the potential energy distribution for a positive ion is like that in fig. 3.19 *a*, so that the energy of the ion is the same whether it occupies site a or b. If initially it is in site *a* and we apply a positive electric field, provided ϕ is not too large compared with kT, the ion will eventually jump into site b. However, if the strength of the field is increasing with time, ΔW may become quite large by the time this particular ion jumps from a to b. The jumping ion will be accelerated by the applied field : it will acquire extra kinetic energy $(= \Delta W)$ above its normal thermal energy but it will lose this extra energy by collision with the other ions surrounding site b. Thus when ion jumping lags behind the electric field, energy is gained by the ion from the field and then converted into heat in the surrounding lattice or network. We can also see why there is a peak in the energy loss (equation (3.30) and fig. 3.17) with increasing frequency. Suppose we have a large number of ions trapped in separate, identical, double wells like that in fig. 3.19. The total energy loss will depend upon the number of ions undergoing delayed jumps and on the average value of ΔW. Now in an alternating applied field, the rate of change of field increases with the frequency as well as with the amplitude, i.e. if

$$E = E_0 \cos \omega t$$
$$dE/dt = -\omega E_0 \sin \omega t.$$

But the greater the rate of change of field the greater, on average, the value of ΔW. Thus ΔW increases as the frequency increases. On the other hand, at high frequencies very few of the ions will jump before the field is reversed so that while ΔW may increase, the number of ions jumping decreases. At low frequencies the energy loss is small because ΔW is small, while at high frequencies the loss is small because the number of ions jumping is small.

In glasses, many more ions can contribute to dielectric loss than can contribute to d.c. conductivity. Dielectric loss need only involve ions jumping between adjacent sites like a and b, or c and d in fig. 3.10, but steady-state d.c. conduction must involve ion transport right through the glass and therefore is equivalent to ions jumping from site a through to site d.

Since for an ion jumping mechanism we expect τ to be related to the activation energy for jumping then, according to the random network model for the structure of glass, we must also expect that there will be a wide range of values for τ corresponding to the range of individual barriers. This is what spreads the dielectric losses and the changes in permittivity over a wide range of frequency : a series of energy loss peaks, and a series of closely spaced steps in permittivity, like the single ones in fig. 3.17, from frequencies $\sim 10^{10}$ to $\sim 10^{-3}$ Hz must be superposed to produce the variation observed, cf. fig. 3.18.

3.13. *Dielectric absorption currents*

When a d.c. potential is applied to a dielectric, charges are displaced within the material. Moving charges constitute an electric current so that the polarization of a dielectric between the plates of a capacitor produces a flow of current in the circuit ; this current, known as the *dielectric absorption current*, will cease when polarization is complete.

Glasses exhibit relatively large delayed polarization effects. The displacement of significant numbers of charges occurs with very long relaxation times and there is a correspondingly prolonged dielectric absorption current.

Suppose we put a capacitor in series with a battery, switch, resistance and a current meter, as in fig. 3.20. If the capacitor were ideal, it would have an infinite resistance and polarization of the dielectric would be instantaneous, so that on closing the switch, current would flow until the charges built up on the plates created a potential difference equal to the e.m.f. of the battery. This charging current would be given by

$$I = \left(\frac{V}{R}\right) \exp\left(-t/RC\right).$$

It falls virtually to zero when $t \sim 10\ RC$, e.g. for $R \sim 1\ \mathrm{M\Omega}$ and $C \sim 10^{-9}$ F, in a time of the order of 10^{-2} s.

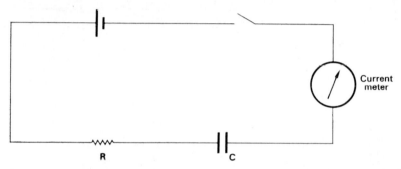

Fig. 3.20. Measurement of capacitor charging current.

If the capacitor has a glass as the dielectric, there is a high initial charging current but this dies away very much more slowly than for an ideal capacitor and eventually reaches a very small steady value. The slower fall in the charging current is due to the delayed polarization in the glass and the small steady current is due to the finite conductivity. The slow decrease in current corresponds to the 'transient' current referred to earlier in the discussion of the d.c. conductivity of glass.

When a glass-filled capacitor is discharged, a slow decay in the discharge current is also observed which corresponds to the slow relaxation of polarization in the dielectric. These slow changes in the charging and discharging currents are illustrated in fig. 3.21. If we short-circuit a fully charged capacitor and then open the circuit again after a short interval of time $(t > RC)$, the charge on the plates of the capacitor will slowly build up again as the delayed polarization in the dielectric relaxes. This effect provides a convenient and simple way of studying delayed polarization; the charge built up on the capacitor can be measured after various time intervals.

83

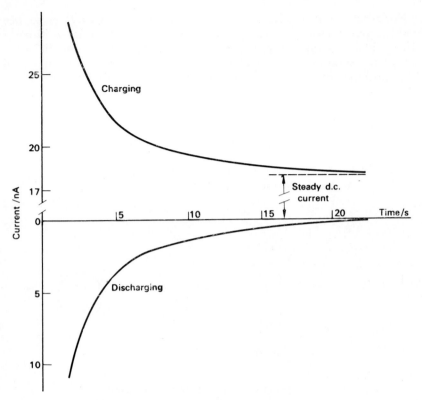

Fig. 3.21.   Dielectric absorption currents on charging and discharging a glass-filled capacitor.

Dielectric memory—the principle of superposition

The slow relaxation of delayed polarization provides the dielectric with a short-term ' memory '.   The polarization at any instant depends upon previously ' experienced ' electric fields.   This can be demonstrated in a rather striking way by applying in succession two fields of opposite polarity and then measuring the dielectric absorption current arising from the subsequent relaxation of the polarization. Figure 3.22 illustrates the result of such an experiment carried out in our own laboratories using an ordinary jam-jar (soda–lime–silica glass). Aluminium foil electrodes were fastened to the bottom half of the outside and inside surfaces of the jar to make a capacitor.   Then with the outer electrode earthed, we applied $+2.5$ kV to the inner electrode for 30 s followed immediately by -1.5 kV for 15 s.   The capacitor was then short-circuited for about 1 s, to remove both the charges on the electrodes and the instantaneous polarization in the glass, and then con-

84

nected to a sensitive current meter in order to follow the discharge current due to the decay in polarization with time. As is shown in the figure, the discharge current falls quite quickly to zero but then reverses direction and rises to a maximum in the opposite direction before finally dying out altogether !

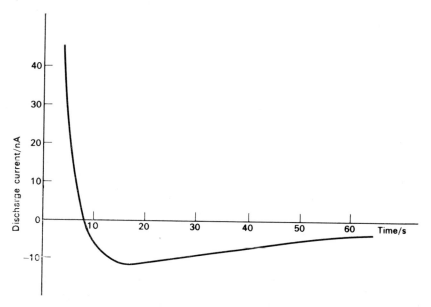

Fig. 3.22. Reversal of discharge current from a glass-filled capacitor, demonstrating the memory effect.

Delayed polarization effects are quite pronounced in glass, taking place in a time-scale which is readily accessible, and their existence has been known for a very long time. Benjamin Franklin described some experiments on the repeated discharge of a Leyden jar, in 1748, and J. Hopkinson carried out an extensive series of experiments between about 1880 and 1890 from which he inferred that the absorption currents in glass obey the principle of superposition. This principle is that sequential changes in voltage produce currents which are additive. A simple example will help to illustrate the basic idea. Suppose we apply a voltage V_1 at time $t=0$ to a sample of glass and measure the current $(=I_1)$ at a time t_3 later ; then in a completely separate experiment we apply a voltage V_2 at time t_2 and measure the current $(=I_2)$ at time t_3 (i.e. (t_3-t_2) after applying the voltage). The principle of superposition asserts that if, in a third experiment, we apply a voltage V_1 at $t=0$, increase it to (V_1+V_2) at time $t=t_2$, then the current flowing at $t = t_3$ will be $(I_1 + I_2)$. Consideration of the special case where $V_1 = V_2$ and $t_2 = 0$ so that $I_1 = I_2$ serves to illustrate a further implication of the principle, that the absorption current after any given interval is proportional to the applied voltage.

The principle of superposition also implies that the initial change in current with time following the removal of an applied voltage is identical to the change which occurred following the application of that voltage. The curves in fig. 3.21 can be superposed after subtracting the steady d.c. component from the charging current curve by reversing the sign of one of the currents.

Qualitative explanation of current reversal

It was suggested earlier in this chapter that dielectric absorption currents are due to the delayed movement of the positively charged alkali metal ions. Since there are positively charged carriers only, the reversal of current illustrated in fig. 3.22 must correspond to a *net* flow of charges first in one direction and then in the other. At first sight this may seem unlikely but it is in fact possible to account quantitatively for the changes in current using the principle of superposition and the theory of relaxation processes outlined earlier. We can see, qualitatively, how the reversal of current comes about by recalling that some of the alkali ions require only very small activation energies to jump into neighbouring holes in the network, while others require much higher energies. In an applied electric field, the ions ' facing ' lower barriers will on the average jump in the field direction, and thus contribute to the absorption current, after a relatively shorter time. In order to simplify the argument, suppose we have just two groups of ions, group A ions having a much lower activation energy than group B ions. In an electric field which has been applied for some time, large numbers of both A and B ions will be displaced ; if the field is reversed group A ions will move to new positions relatively quickly but, since the probability of group B ions jumping is much smaller, these ions will retain their displacements in the direction of the first field for a longer time. Suppose the reverse field is removed before any significant number of group B ions have jumped back under the reverse field, then the decaying absorption current observed initially will be due to the movement of group A ions which will return to their normal sites relatively quickly. Subsequently, a smaller current in the opposite direction will be observed due to the movement of the more sluggish group B ions still returning from the displacement suffered under the first applied field. If the relaxation times for the displacement of ions in each group are τ_A and τ_B respectively, then to obtain an impressive reversal of current we need $\tau_A \ll \tau_B$, the first field must be applied for a time $\gg \tau_B$ and the reverse field must be applied for a time $\gg \tau_A$ but $\ll \tau_B$.

86

Dielectric absorption currents in glasses are produced by alkali metal ions which can move through the network but have rather long relaxation times and eventually become trapped in what we might imagine as a ' blind canyon ' of potential barriers. There is a whole spectrum of relaxation times at any given temperature. The distinction between dielectric losses and dielectric absorption currents is not fundamental but is a matter of convention ; it springs from the fact that the same process may produce an electrical effect which can be detected by either a.c. or d.c. techniques depending on the relaxation time. A dielectric absorption current can be considered as a dielectric loss at very low frequencies, say, < 1 cycle per minute. When the temperature is raised the relaxation times will fall ($\tau \propto \exp \phi/kT$) and losses due to the displacement of the same ions will occur at much higher frequencies.

If the longest relaxation times are only $\sim 10^{-1}$ s then the effects will all appear in the form conventionally regarded as ' dielectric loss ' and there will be no ' absorption currents '.

CHAPTER 4

optical properties of glass

THE most obvious optical property of ordinary silicate glasses is their very high transparency to visible light.   It is this characteristic, combined with hardness and chemical stability, which has led to the demand for vast quantities of glass for use as containers and windows of all kinds and thus to the development of the major manufacturing industry. The smaller, but highly specialized, technical industry concerned with the melting and processing of a wide variety of glasses for use in lenses and other optical devices is of more recent origin.

The optical properties of solids are not directly dependent on the form or even existence of long-range order but, rather, on the nature and local arrangement of the constituent atoms or ions.   The optical properties of glasses are, therefore, not sharply distinguished from those of crystals which contain the same ions in comparable local groupings, although the variety of ions which can be incorporated into a silica network does yield a remarkable flexibility in the refractive index and the colour of a silica-based glass.

In this chapter we shall be concerned first of all with the basic problem of the interaction between light and matter in bulk ; what is it that determines the refractive index of a material, and how ?   Then in the later sections, we shall describe the optical characteristics of glasses and the methods of producing optical and coloured glasses.

4.1. *Transparency of glasses*

For a block of material to be transparent two conditions must be satisfied.   The material must be homogeneous at least down to a scale comparable with the wavelength of light, and it must have a low absorption at least for some wavelengths.   If, within the block, there are regions which are large compared with the wavelength of light and which have refractive indices significantly different one from another, then refraction or internal reflection will occur at some of the boundaries between these regions.   This occurs, e.g., in milk, plaster of paris, snow, etc.   Thin sections of such materials may be translucent if the number of boundaries within the thickness is small ; some light will pass right through the section, although bunches of parallel rays will be broken up and it will be impossible to form an image on one side of the material of an object on the other.   In a literal sense, therefore, the material is not transparent.   A single irregular boundary can serve to destroy the transparency of a block of material ; a plate of glass with ground surfaces appears translucent rather than transparent.

Opaque glasses can be produced and are in common use for decorative tableware, wall-cladding and light-fittings. These glasses contain small particles which have a different refractive index from that of the matrix glass. There are two main types : crystalline opals and emulsion opals. The crystalline opals are produced by selecting a glass composition from which small crystals, e.g. of calcium fluoride, will precipitate as the melt cools. The emulsion opals contain small ' droplets ' of a glass based on P_2O_5 as a network former dispersed in a silica glass ; they are analogous to the liquid emulsions which can be formed from two immiscible liquids such as oil and water.

The distinction between transparency and translucency (or opacity) resulting from the scattering of light by external or internal boundaries, although important in practice, is trivial from a fundamental point of view. Of much greater interest is the fact that all materials extract energy from a beam of light to a greater or lesser degree. For microscopically homogeneous materials it is found experimentally that the light intensity within the material falls off exponentially with distance :

$$I = I_0 \exp\left(-\beta_T d\right) \quad \text{(Lambert's law)}. \tag{4.1}$$

I_0 is the initial intensity, d is the distance and β_T is a property of the material and is called the *absorption coefficient*. The fundamental condition for transparency is that β_T should be small and if its magnitude varies significantly with the wavelength of light in the visible range the material will be coloured.

For practical purposes it is often more convenient to work in terms of the overall *transmittance*, T, rather than the absorption coefficient. The transmittance is defined as the ratio of the intensities of the emergent, I_e, and incident, I_i, beams of light. Thus for a block of material of thickness t :

$$I_e/I_i = T = (1-R)^2 \exp\left(-\beta_T t\right), \tag{4.2}$$

where R is the fraction reflected at each interface. For near-normal incidence $R = (n-1)^2/(n+1)^2$, n being the ratio of the refractive indices of the block and of the surrounding medium. The transmittance is not a material property since it depends on the thickness of the sample. Typically for optical quality glass, β_T is of order 1 m^{-1} over the whole of the visible spectrum and therefore about 1 per cent of the light intensity is lost for each centimetre of glass traversed. Hence for a glass plate 1 cm thick with $n = 1.5$, $T = 0.91$.

Glasses with even smaller absorption coefficients can be produced and there is a growing demand for these. Bundles of fibres can be used as a ' light-pipe ' ; light entering a fibre at one end suffers total internal reflections at the longitudinal surfaces and can emerge only at

89

the far end of the fibre.   Light propagates along the fibre even though
the fibre is bent (fig. 4.1), provided the radius of curvature is greater
than a critical value for a given diameter of fibre, so that fibres can be
used as a flexible light-guide.   The use of a bundle of fibres rather
than a solid rod not only confers flexibility and decreases the optically
critical curvature, but also permits the transmission of an image from
one end to the other.   The grain size in the transmitted image will be
equal to the diameter of the individual fibres.   If the fibres in the
bundle are not aligned and each fibre occupies a different relative
position at the two ends, the bundle will serve as an image ' scrambler ',
a photographic record of an image can be decoded only by the same
fibre bundle.   The practical limit to the useful length of such *fibre-
optics* devices is set by the absorption in the glass.

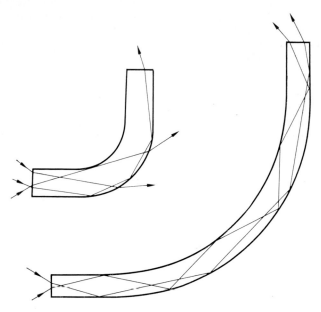

Fig. 4.1.   The paths of light-rays along curved glass rods.

4.2. *Absorption and scattering processes*

Two distinct effects contribute to the value of the absorption
coefficient for a given material : a *true* absorption in which the energy
of the light is converted into heat, and *scattering* by particles which are
much smaller than the wavelength of the light.   Both of these processes
cause an exponential decrease in the intensity of the light with distance,
and β_{T} in equation (4.1) is given by $(\beta_{\mathrm{a}} + \beta_{\mathrm{s}})$, where β_{a} is due to true
absorption and β_{s} is due to scattering.

Scattering of light from an incident beam results in a finite intensity at an angle to the beam, at the expense of the light transmitted in the forward direction. For scattering by particles which are smaller than the wavelength of light, the intensity of the scattered light is proportional to the square of the particle volume and to λ^{-4}, so that short wavelengths are much more strongly scattered than long ones. The ratio of the intensities of incident to scattered light will be ~ 10 times greater for violet light, $\lambda \sim 400$ nm, than for red light, $\lambda \sim 720$ nm. It is this process, known as *Rayleigh scattering*, which accounts for both the visibility and the bluish colour of tobacco smoke. In spite of the fact that the individual smoke particles are much smaller than the wavelengths of visible light, such smoke clouds are visible because the intensity of scattered light is high and they are 'blue' because the intensity of the scattered light is much greater for the shorter wavelengths.

Rayleigh scattering becomes very small for particles of atomic dimensions but it is still observable. It is the scattering by small groups of molecules, formed by temporary local fluctuation in density, which is responsible for the blue colour of the sky and for the orange colour of the setting Sun ; in the absence of a thick layer of air, e.g. from the surface of the Moon or even from a high-flying aircraft, the sky overhead appears black.

On the other hand, when the size of the particles becomes comparable with the wavelengths of light, diffraction effects and reflection from the surfaces of the particles, as well as refraction and absorption within individual particles can all occur. Under these conditions, the intensities of the scattered and of the transmitted light may show a quite different variation with wavelength but the exact form depends in a very complex way on the optical properties of individual particles and on their size. Scattering by particles of volcanic ash in the atmosphere has been responsible for the occasional *blue* Moon.

4.3. *The origin of refractive index and dispersion*

Although many of the effects associated with the refraction of light at the boundary between two media are treated in detail in introductory physics courses, the physical processes ultimately responsible for these effects are probably not familiar. Even with a highly transparent material, light does not 'just pass through' ; there is an intimate interaction between the light and the charged particles which make up the material.

Light is sometimes described as an electromagnetic wave and sometimes as a stream of particles (photons). Each photon in a beam of monochromatic light of frequency v has energy hv and momentum hv/c. Despite the superficial contrast, these two descriptions do not seriously conflict, rather, they are considered to be complementary. They

91

reflect the characteristics of light which have been established by experiment ; for example, the wave nature revealed by interference phenomena and the particle-like behaviour apparent in the photoelectric effect. In modern physics, for a descriptive account of the behaviour of light, it is sometimes simpler to use the wave picture, sometimes the particle picture. The formal quantum theory of light allows one to predict whether the wave or the particle nature will be dominant in any given situation.

Many of the optical properties of materials can be qualitatively explained using the wave theory. Refraction of a light beam at the boundary between two media occurs because the speed of the waves is different in different media ; the refractive index is just the ratio (speed of light waves in vacuum)/(speed in the material). Dispersion is the variation of the speed of light waves with frequency ; a glass prism produces an angular separation of the component waves in white light from a slit, only because the refractive index is different for different wavelengths. A ray of blue light is deviated by a prism more than a ray of red light because the refractive index for glass, and indeed for most materials, increases with frequency in the visible part of the spectrum. Clearly, the basic questions here are—why does the speed of light depend on the medium ? And, why does it increase with the frequency ?

A formal, although perhaps not very satisfying, answer can be extracted from Maxwell's electromagnetic theory, in which the transverse vibrations in a light wave are identified with fluctuating electric and magnetic fields. It follows from the mathematical theory that the speed of propagation of the waves depends on the electric and magnetic properties of the medium. As we have already mentioned, the refractive index is given by

$$n = c/c_{\mathrm{med}} = \sqrt{(\mu_{\mathrm{r}}\epsilon_{\mathrm{r}})} = \sqrt{\epsilon_{\mathrm{r}}} \quad \text{when} \quad \mu_{\mathrm{r}} = 1$$

where μ_{r} and ϵ_{r} are the relative permeability and relative permittivity, respectively. Thus the speed of light in a material depends on its dielectric polarizability. In order to describe the physical basis of this dependence and thus give a more descriptive answer to the questions above, we must take account of the nature of optical media as well as the nature of light.

All media, other than the idealized empty space, contain charged particles, possibly ions but certainly electrons and nuclei. Many of these particles can be displaced from their normal positions against restoring forces approximately proportional to the displacement. A physical explanation of the variation of the speed of light can be given by considering light to be electromagnetic waves and materials to be collections of charged harmonic oscillators.

When a beam of light passes into a material, the transverse alternating

92

electric field of the electromagnetic wave excites forced oscillations of the charged particles. As we shall describe in more detail in the following sections, the speed of the light waves inside the material depends upon the amplitude of oscillation and the number density of these charged particles. The oscillation of the charges in any local region of the material produces an alternating dielectric polarization and the amplitude of this is related to the relative permittivity at the frequency of the alternating electric field. Both the refractive index and the relative permittivity are therefore determined by the same physical process, namely the displacement of charges in the material by an applied electric field ; this is the physical reason for the quantitative relationship between them.

We shall use this equation, together with the properties of harmonic oscillators, to derive an expression for the refractive index of a model material in terms of the number and resonant frequency of the charged particles, but first of all we shall outline qualitatively the optical effects produced by the forced oscillations of the particles.

Oscillating charges and the speed of light

An oscillating charge is equivalent to an alternating current in a miniature aerial : it radiates electromagnetic waves in all directions (except, in the case of a charge oscillating on a straight line, parallel to the line of oscillation), and the frequency of the radiation is the same as that of the oscillations. If the charges in a material behave as harmonic oscillators, then the forced oscillations induced by the electric field of a light wave will have the same frequency as the original wave. Each oscillating charge in a material acts therefore as a *real* secondary source of waves of the same frequency as the incident light, and the original electromagnetic wave is modified by the superposition of the secondary waves emanating from all the oscillating charges. According to this physical model, the medium simply provides an array of secondary sources and the waves from these interfere with one another, and with the primary wave, to produce new resultant waves. The primary and secondary waves all travel with the speed of light in vacuum. A detailed analysis of these interference effects is rather involved and beyond our scope here ; we shall just summarize some of the results.

For simplicity let us suppose that we have a parallel beam of light, i.e. electromagnetic waves with plane wave fronts, incident normally on the surface of a material. Although each oscillating charge in the material radiates in all directions, this does not necessarily lead to resultant waves in all directions. In fact if the charges are regularly arranged in space and are close together compared with the wavelength, then the phases of the secondary waves are such that they will interfere destructively and completely except in the direction of propagation of the main beam. In this respect the behaviour of these real secondary

93

waves is similar to that of the *virtual* waves radiated by the *imaginary* secondary sources on Huygens' wave theory. When the charges are not regularly arranged in space (e.g. in a gas), or when a regular array is confined to a region smaller than the wavelength of the light (e.g. in a small smoke particle), then destructive interference is not complete and there will be a resultant wave of small amplitude lateral to the main beam ; this corresponds to Rayleigh scattering.

In directions parallel to the main beam, the secondary waves do not completely cancel even when the charges are closely spaced and regularly arranged. A small resultant wave is produced outside the material travelling in the backward direction : this corresponds to the light reflected from the incident surface. In the forward direction, super-position of the primary (incident) wave and the secondary waves, within the material, gives a resultant which has an amplitude similar to that of the primary wave but a different wavelength and speed. The ratio of the wavelengths and the ratio of the speeds are the same, so that we still have, as we should expect, *speed = frequency × wavelength*. The magnitude of this ratio depends on the amplitude of the secondary waves relative to that of the primary wave and hence on the ratio of the amplitude of oscillation of the charges to that of the driving field ; this in turn is simply related to the high-frequency polarizability of the medium. For all materials, it is the electrons which undergo the largest forced oscillations at frequencies corresponding to visible light (compare equation (4.7) below). The optical refractive index of a material is therefore determined by the way in which the electrons are deployed within it.

Oscillating charges and the absorption of light
During its passage through a material, the intensity of light is reduced progressively (equation (4.1)), not only because some of the light is scattered out of the main beam but also because some of the energy is absorbed by the material and converted into heat. This true absorption of light can also be accounted for in terms of the behaviour of oscillating electrons (in the case of visible light) within a material. Occasionally the vibrational energy associated with the forced oscillations of an electron may be converted into heat, that is into energy associated with random atomic vibrations. Once this has happened, energy is extracted from the electromagnetic wave in restoring the normal oscillation amplitude of the electron. The oscillation energy of an electron can be transferred to atomic vibrations when the atom to which the electron is bound collides with a neighbouring atom. In some of these collisions the total kinetic energy of the two atoms immediately after the collision is greater than it was before ; during the collision, the amplitude of oscillation of the electron is reduced and the vibrational energy gained by the atoms is equal to that lost by the electron. This

94

is called an *inelastic collision*. The amount of energy converted into heat by these inelastic collisions will increase as the amplitude of the forced oscillations of the electrons increases : the bigger the energy of the electron oscillations the greater is the possible increase in the kinetic energy of the atoms. Collision losses of this kind result in a true absorption of the light, but they are quite small except when the energy of oscillation of the electrons is large, i.e. they are small except when the frequency of the light is near the natural frequency of free oscillation of the electrons. Therefore, pronounced optical absorption occurs near the resonant frequencies of the electrons in a material.

Quantitative model for refractive index

Using Maxwell's relation $n^2 = \epsilon_r$ and a harmonic oscillator model to represent the forced vibrations of the charges within a material, we can obtain an equation for the refractive index, in terms of the number and properties of the charges, which represents the main physical features of dispersion. We shall now outline the calculation for the simplified situation in which the damping of the oscillators (i.e. the energy loss by collision and by re-radiation) is sufficiently small that it may be neglected.

The particles, of mass m and charge q, are subjected to a periodic electric field, which we can represent by $E = E_0 \cos \omega t$, and experience a restoring force proportional to the displacement, x, from their normal positions. The equation of motion of one of these particles is, therefore

$$m \frac{d^2 x}{dt^2} + \beta x = E_0 q \cos \omega t, \qquad (4.3)$$

where β is the restoring force per unit displacement. The steady-state solution of this equation is

$$x = \frac{E_0 q \cos \omega t}{(\beta - m\omega^2)} = \frac{E_0 q}{m(\omega_0^2 - \omega^2)} \cos \omega t, \qquad (4.4)$$

where $\omega_0 = (\beta/m)^{1/2}$ is the natural frequency of free oscillations. Thus the oscillations have the same frequency as the driving field and their amplitude depends on the frequency of this field. The change in sign of x which occurs as ω passes through ω_0, corresponds to a change in the phase of the oscillations with respect to the driving field.

Any harmonic oscillator will exhibit this effect. It can be quite simply demonstrated using two simple pendulums suspended from a common support. If we use a much larger mass for the bob of one pendulum than for the other, then when the heavier pendulum is set into oscillation it will drive the lighter one. The phase difference between the two oscillating pendulums depends on the relative lengths of the suspending strings, i.e. on their natural frequencies. At resonance ($\omega = \omega_0$) the phase difference is $\pi/2$ and the amplitude of the

95

lighter pendulum remains finite, due to the damping. The greater the damping of a forced harmonic oscillator, the smaller is the amplitude at resonance and the larger is the frequency range over which the phase change $0 \to \pi$ takes place (fig. 4.2).

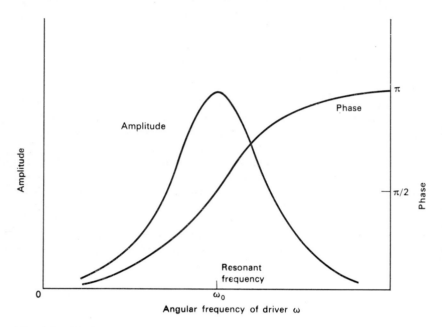

Fig. 4.2. Variation with driving frequency of the amplitude and phase of a forced oscillator of natural frequency ω_0

The displacement of the charged particles from their normal positions in a material produces induced dipole moments xq for each particle (*dipole moment = charge × displacement*). The local dielectric polarization, P, (dipole moment per unit volume) will be $n_0 xq$, where n_0 is the number of dipoles per unit volume.

But, from equation (3.10)

$$P = \epsilon_0(\epsilon_r - 1)E.$$

If n_0 is small, so that there is negligible interaction between dipoles, then the field inducing any dipole is equal to the applied field and

$$P = n_0 \alpha E$$

and

$$(\epsilon_r - 1) = n_0 \alpha / \epsilon_0,$$

96

where α is the polarizability of the charge $= \dfrac{\text{induced dipole moment}}{\text{applied field}}$

$= xq/E_0 \cos \omega t$. Using Maxwell's relation and equation (4.4) we have

$$(n^2 - 1) = (\epsilon_r - 1) = \frac{n_0 \alpha}{\epsilon_0} = \frac{n_0 q^2}{\epsilon_0 m(\omega_0{}^2 - \omega^2)} \tag{4.5}$$

but $n \simeq 1$ if n_0 is small, and then $(n^2 - 1) = (n - 1)(n + 1) = 2(n - 1)$ so that

$$n = 1 + \frac{n_0 q^2}{2m\epsilon_0(\omega_0{}^2 - \omega^2)}. \tag{4.6}$$

The variation of n with frequency $(\omega/2\pi)$, according to this equation, is shown in fig. 4.3. The curve for finite damping is also illustrated.

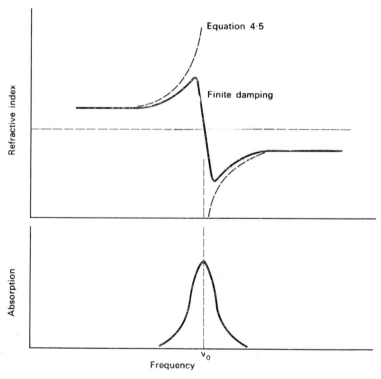

Fig. 4.3. Variation of refractive index near an absorption band.

If n_0 is not small, then the equation $P = n_0 E$ is not valid, because the internal field is not equal to the applied field. However, we can calculate ϵ_r, and hence n^2, by using the Clausius–Mossotti equation

97

(equation (3.24)). This just involves replacing the left-hand side of equation (4.5) by

$$3\frac{(n^2-1)}{(n^2+2)} = 3\frac{(\epsilon_r-1)}{(\epsilon_r+2)} = \frac{n_0\alpha}{\epsilon_0}.$$

In any real material there will be charged particles with different values of m and β: nuclei, or ions, and electrons have very different masses and there may also be several groups of electrons each with a different value of β, the restoring force constant. For real materials, therefore, there will be a number of different resonant frequencies and the right-hand side of equation (4.5) is replaced by a series of separate terms, one for each group of charges. We have then

$$\frac{(n^2-1)}{(n^2+2)} = \sum \frac{n_0}{3\epsilon_0 m}\left(\frac{q^2}{\omega_n^2-\omega^2}\right) \qquad (4.7)$$

where there are n_0 charges per unit volume corresponding to each resonant frequency, ω_n.

If we allow for the damping of all the oscillators, then the refractive index will not rise to infinity whenever the frequency corresponds to one of the resonance frequencies, although we shall expect the absorption to increase sharply near each resonant frequency.

There is nothing in our analysis which limits the frequency of the radiation to the visible range. We can expect therefore that the actual variation of refractive index and absorption coefficient over a wide frequency range might be something like that in fig. 4.4. Because of

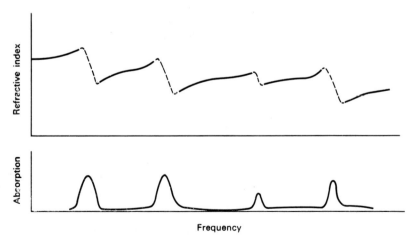

Fig. 4.4. Expected variation of refractive index and absorption over a wide range of frequency.

98

the large difference between the masses of nuclei, or ions, and of electrons, the terms in equation (4.7) due to the oscillations of the heavier particles will be $\sim 10^{-4}$ times those due to electrons, for similar values of $(\omega_n{}^2 - \omega^2)$. This is the reason why the optical properties of a material depend primarily on the arrangement of the electrons.

We have followed the arguments leading to the dispersion equation using a wave picture for light and assuming that the charged particles behave classically. It may be interesting to note that the quantum theory yields an expression for n which is of exactly the same *form*, although the meaning of some of the parameters which occur in the equation is different. For example, the resonant frequencies of the classical picture become energy level transitions in the quantum theory, and the number of oscillators corresponding to each resonant frequency becomes a transition probability.

It is not yet possible to predict theoretically, or even to identify from experimental observations, all the individual resonant frequencies (or quantum level transitions) for charges which exist in a simple solid let alone a complex glass. Nevertheless, the physical picture provided by the classical model is quite useful. The transparency of many ionic crystals and of glasses over the whole of the visible spectrum indicates that the resonant frequencies of all the charges in these materials lie outside this frequency range. Strong absorption does occur in ionic crystals and in glasses in both the infra-red and the ultra-violet (fig. 4.5),

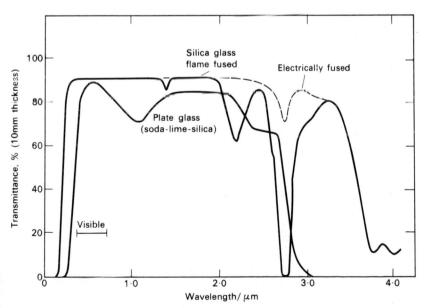

Fig. 4.5.   Transparency of some commercial glasses.

99

and is due to the resonant vibrations of ions and electrons, respectively. Dispersion in these materials, and its marked increase towards the violet end of the spectrum (fig. 4.6), is associated with the ultra-violet absorption bands due to the electrons attached to ions in the structure.

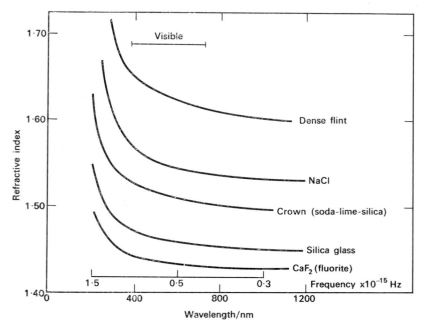

Fig. 4.6.   Dispersion in some ionic crystals and glasses.

Many crystals and some kinds of glass owe their colour to a broad absorption band which covers part of the visible frequency range, and we shall consider these in section 4.7.

Infra-red absorption
The absorption bands in the infra-red, due to the resonant vibrations of the ions in glasses, are of intrinsic interest for the information they may reveal about the structure of the glass.   It is, obviously, difficult to measure the refractive index or the absorption coefficient, in transmission, near a very strong absorption band.   However, the position and shape of the peak of the absorption band can be found by examining the radiation reflected from the surface of a material, as a function of frequency.   Resonance of the oscillators near the surface of the material gives rise to a more pronounced backward wave.   This effect is more immediately obvious in the visible part of the spectrum.   Materials

which owe their colour to a very strong selective absorption, reflect the complementary colour ; very thin films of gold are transparent and appear blue-green in transmitted light but are the familiar reddish-yellow by reflection.

P. E. Jellyman and J. P. Proctor have recently studied the infra-red reflectivity of several series of simple glasses and, by progressively varying the composition, were able to identify some of the observed bands with the resonances of particular ion pairs. Figure 4.7 shows some of their results and also some of the pronounced reflection peaks which are observed with ionic crystals due to the resonant vibrations of the ions on the lattice.

For the glasses the peak at 9 to 9·5 μm is attributed to a bond-stretching vibration between the Si and O ions in a Si–O–Si linkage and the second peak at 10–11 μm to bond-stretching in a Si–non-bridging oxygen linkage. At first sight these observations may appear to conflict with the results for silica, illustrated in fig. 4.5, which show

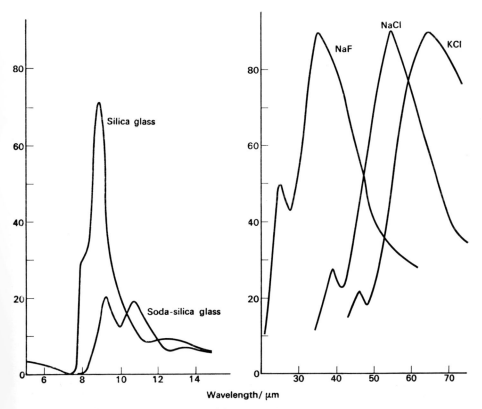

Fig. 4.7.   Reflectivity in the infra-red for some simple glasses and ionic crystals.

that flame-fused silica has a strong absorption band at about 2·7 μm. However this band is due to vibrations of Si–OH bonds formed when water is introduced into the network during fusion. The transmission curves of fig. 4.5 are for plates of normal thickness (∼ 10 mm), so that quite small concentrations of hydroxyl ions will produce significant absorption. The small bands at ∼ 2·2 and 1·4 μm are also associated with the 'water' present as an impurity, and electrically fused silica has a much more uniform transmission out to ∼ 4 μm. This final cut-off is due to the edge of the extremely strong absorption of the Si–O network.

4.4. *Refractive index and dispersion of glasses*

Using the classical model described in the previous section we can predict that a higher refractive index will be associated with a greater electron density and thus with higher mass density and/or higher atomic number. This is observed in practice as is illustrated, for example, by the increase in the refractive index of a gas with its density and by the examples in table 4.1. NaCl and NaF have similar mass densities but

	Relative density	Refractive index
Forms of SiO$_2$: Quartz	2·65	1·55
Cristobalite	2·32	1·49
Tridymite	2·28	1·47
Silica glass	2·20	1·46
NaF	2·6	1·34
NaCl	2·2	1·54
CaF$_2$	3·18	1·43
CaCl$_2$	2·51	1·52

Table 4.1. Comparison of refractive indices.

chlorine has a higher atomic number than fluorine, and therefore more electrons, so that NaCl has more electrons per unit volume than NaF and consequently a greater refractive index. Silica can exist in three different crystalline forms as well as the glassy form ; the densities of these structures are different and the refractive index increases when the SiO$_4$ tetrahedra are more closely packed. For a glass of given composition we have seen that the density depends on the rate at which it is cooled through the transformation region. Not surprisingly, therefore the refractive index also depends on the rate of cooling. Although the change in refractive index is quite small, affecting the third decimal place, it can be very important ; precise optical components require very careful annealing to ensure that a reproducible index is obtained which is uniform through the thickness of the glass.   Among the silica-based glasses, those containing higher proportions of heavy ions such as Pb or Ba exhibit larger refractive indices.   Indeed, it was discovered experimentally many years ago that, as we might expect from the model, the refractive index of glass is an additive property (see section 2.2) ; using previously determined factors, the glass technologist can estimate to within a few per cent the effect of a change in composition on the refractive index.

A number of empirical equations have also been proposed to describe the variation of refractive index with wavelength over the visible range, the constants in the equations being determined by experiment for any particular glass.   The more elaborate of these equations take a form similar to equation (4.7) and once the constants have been evaluated from measured refractive indices at selected frequencies, they can be used to calculate the value of the index at some intermediate frequency to an accuracy of 1 part in 10^5.   Such high accuracy is essential in the design of modern multicomponent lenses.   The simplest dispersion equation is that due to Cauchy

$$n = A + B/\lambda^2 + C/\lambda^4,$$

where for a soda–lime–silica glass A, B and C are of the order of 1·4, 10^{-14} m^2 and 10^{-28} m^4 respectively.   This equation represents fairly well the variation of refractive index for normal commercial glasses, and for ordinary purposes (to within ~ 1 per cent, or the third decimal place) we can omit the third term (C/λ^4).

4.5. *Optical glasses*

Although the use of glass to make lenses goes back to before the time of Newton and Galileo, and Dolland was awarded a patent in 1758 covering the invention of an achromatic doublet lens, it was not until the beginning of the nineteenth century that serious attempts were made to produce, commercially, glass suitable for the manufacture of lenses or other parts of optical instruments.   Not until the second half of the century was there a variety of ' optical ' glasses available with a useful range of refractive indices and dispersions.

Whatever the composition, glasses of a quality suitable for use in optical instruments must be produced by special procedures to ensure a very high homogeneity and thus a highly uniform refractive index. Ordinary glass, which has been melted in bulk in tank-furnaces or large pots, such as is used to produce glass plate and containers or tableware, invariably has small variations in composition from one region to another.   These inhomogeneities in composition become drawn out into layers or streaks, known as ream or striae, as the glass is formed into the product and they give rise to slight variations in refractive index.   In plate glass for example, the layers are roughly parallel to the surface of the glass and can be revealed by viewing a strip of the plate ' end-on ' in transmitted light (fig. 4.8). Light rays passing normally through the thickness of such a plate, and therefore nearly perpendicular to the striae, are not significantly affected by the variation in refractive index. However, if such material were fashioned into a lens, deviations of the oblique rays would seriously affect the quality of an image. The actual variation in refractive index across a typical stria is quite small, $\Delta n/n \gtrsim 0.01/1.5$, and even ' ordinary ' quality glass is substantially homogeneous. Until quite recently the high homogeneity required for optical components was achieved by melting the glass in relatively small pots, holding about 1000 kg or less, and the highest quality melts are still carried out in crucibles of

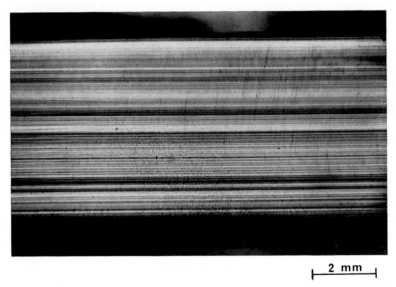

2 mm

Fig. 4.8. Striations in a sample of plate glass, viewed in transmitted light.

platinum alloy in order to minimize the contamination of the glass by the refractory. In the traditional process the glass is stirred after refining and then allowed to cool in the pot ; during cooling, the mass of glass cracks into irregularly shaped lumps and these are sorted to eliminate those which still contain striae or small bubbles. The lumps are reheated and moulded roughly to the shape of a particular optical component and then ground and polished to the final form. Since 1948 a continuous process has been in use for producing optical glass ; stirring is still essential but melting, refining, and mechanical stirring are carried out in successive regions of the same tank, before the glass reaches the coolest section and is brought to a temperature at which it can be moulded or cast into blocks.

The first optical glasses were a soda– or potash–lime–silica and a lead–potash–silica glass ; these were the original *crown* and *flint* glasses, respectively. Flint glass has a higher refractive index and also a higher dispersion than crown glass so that if two prisms of the same refracting angle are made, one from flint and one from crown glass, then both the angle of minimum deviation for any wavelength and the angular spread of the spectrum from white light are greater with the flint glass prism. Unfortunately the spectra produced by the two prisms are not just different in scale ; see fig. 4.9. The dispersions in different parts of the spectrum (the partial dispersions) are not proportional to the total dispersion. This is why the simple achromatic doublet lens can reduce but not eliminate chromatic aberration. The achromatic doublet is

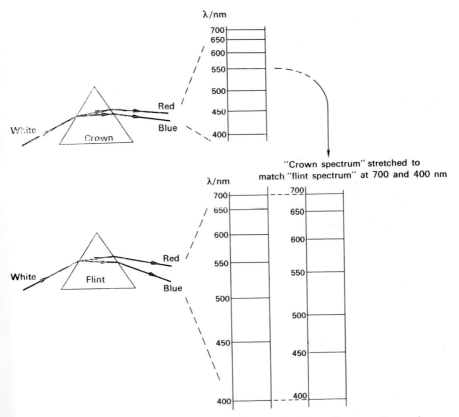

Fig. 4.9. Comparison of the spectra produced by flint and crown glass prisms of the same refracting angle.

105

made by combining a strong converging lens with a weaker diverging lens from a glass with higher total dispersion ; the lenses can be chosen so that parallel rays of any pair of wavelengths are brought to a common focus, fig. 4.10. However, rays of any other wavelength will not focus at the same point, so that there is a small residual chromatic aberration known as the *secondary spectrum.*

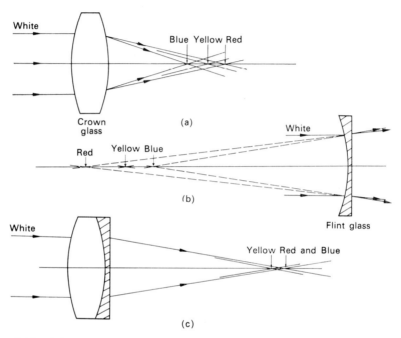

Fig. 4.10. The achromatic doublet. (*a*) and (*b*) show the longitudinal chromatic aberration produced by single lenses (position of focus changes with wavelength) ; (*c*) shows the reduced aberration produced by a doublet lens when the components are chosen so that red and blue rays have a common focal point on the axis.

Many new glasses containing significant proportions of elements from a much wider range than those listed in table 2.2 have been produced in attempting to reduce the secondary spectrum, that is to provide pairs of glasses which have different refractive indices but a more nearly constant ratio of partial dispersions. Ideal pairs do not exist but the very wide selection of optical glasses now available enables lens designers to use three or even more separate components in a compound lens and this helps considerably in the minimization of the secondary spectrum and other image aberrations.

The characteristics of optical glasses are customarily represented by quoting the refractive index for one of the sodium D lines and the constringence, V, which is the reciprocal of the mean dispersive power :

$$V = \frac{n_d - 1}{n_F - n_C},$$

where n_d is the index for the sodium D line 589·6 nm and n_F and n_C are the indices for the hydrogen lines at 486·1 (blue) and 656·3 nm (red) respectively. In catalogues of optical glasses, most manufacturers also

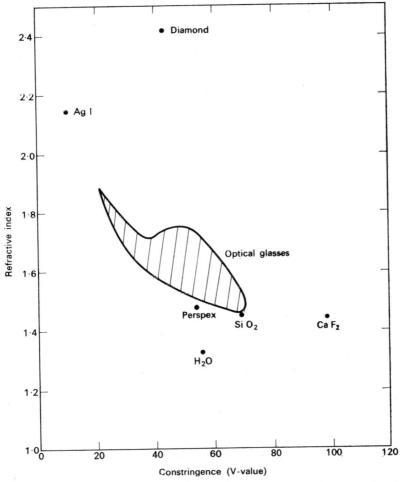

Fig. 4.11. Properties of commercially available optical glasses and other optical materials.

specify the relative partial dispersion, i.e. $(n_{\lambda 1} - n_{\lambda 2})/(n_F - n_C)$, for selected regions of the visible or photographic spectrum. The range of glasses available, illustrated in fig. 4.11, includes those containing Ba, B, F, Zn, Ti and rare-earth ions.

4.6. *Birefringence and the photoelastic effect*

The regular arrangement of atoms, ions or molecules in a crystal makes the structure inherently anisotropic : the sequence and/or, the spacing of atoms is different in different directions within the crystal. The effect may be enhanced by the presence of non-spherical molecules or complex ions, or by low symmetry in the pattern in which the atoms, or etc., are arranged, fig. 4.12. This internal anisotropy is responsible

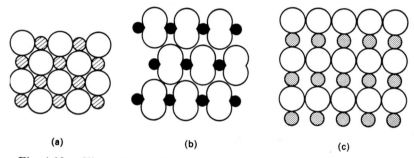

(a)   (b)   (c)

Fig. 4.12.   Illustrating differences in the internal symmetry in crystals.

for a variation, with direction, of some of the physical properties of *single* crystals, although normally for a polycrystalline sample in which the individual crystals are randomly oriented, the effects will average out so that macroscopically it will behave as an isotropic material. The disordered arrangement of the atoms in a glass will ensure that its physical behaviour is isotropic ; there are no directions in which the atomic arrangement is different, on average, from that in any other.

In glass, the speed of light is independent of both the direction of travel and the plane of polarization of the waves, but this is often not the case in a single crystal. In some crystals the electrons can be displaced more easily in one particular direction than in any other so that the dielectric polarizability and consequently the speed of light varies with the direction of the electric field. Light waves which are plane-polarized and travelling so that the electric field vibrations are parallel to the direction of easy electron displacement will have a different speed from those polarized or travelling so that the electric fields are perpendicular to this special one. Such crystals are said to be doubly refracting or *bi-refringent* because when on a face of the crystal a ray of

108

light is incident at an arbitrary angle with respect to the special direction, two refracted rays are produced. The two refracted rays travel through the crystal at different speeds and both are plane polarized, with the directions of polarization at right angles to one another. If the anisotropy is large a complete lateral separation of the two refracted rays occurs, under suitable conditions, so that a double image is formed, fig. 4.13.

More generally, when a sample of an anisotropic material is placed between crossed Polaroids, some light is transmitted by the second (analysing) Polaroid (fig. 4.14) and this may be intensely coloured if a white light source is used.

Many transparent materials, including glasses and plastics, which, ordinarily, are isotropic, become birefringent while subjected to mechanical stress. This is the basis of a very useful experimental technique (called photoelasticity) for determining quantitatively the distribution of stresses in complex engineering structures. It also provides a method for detecting in a glass article, stresses developed by uneven cooling from above the softening point or by a mismatch in the thermal expansion in a glass–metal seal.

If a compressive or tensile load is applied to a glass plate in, say, the y-direction (fig. 4.15) and a beam of light is passed through the plate

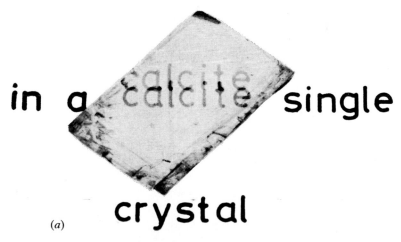

Double refraction

in a calcite single

(a) crystal

Fig. 4.13. (a) Double refraction in a calcite crystal : double image through a single crystal block. Figs 4.13 (b) and (c), overleaf, show the image seen through Polaroid in two positions at right angles, illustrating the polarizations of the two refracted rays.

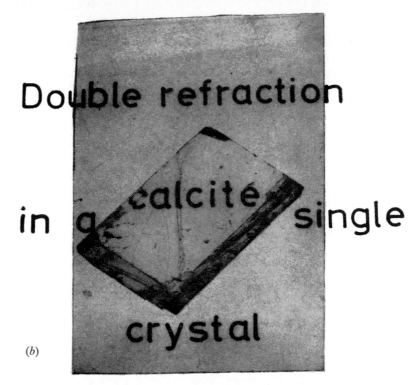

(b)

(c)

Fig. 4.13 (b) and (c).

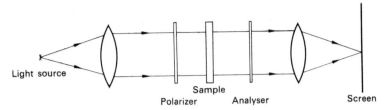

Fig. 4.14. Experimental arrangement for the examination of a sample in polarized light.

in the x-direction, then light waves polarized so that the vibrations are parallel to the y axis will have a different speed from those polarized in the z-direction. Neither speed will be the same as in the absence of stress for there is an elastic strain in both the y and z-directions; the latter is $(\sigma_y/Y)\nu$, where ν is Poisson's ratio. Experiment has shown that the difference between the refractive indices for the waves polarized in the direction of the single applied stress and perpendicular to it is proportional to the stress applied, i.e. if n_y and n_z are the two refractive

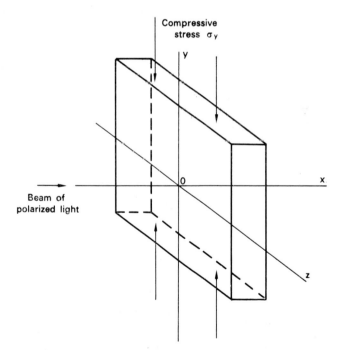

Fig. 4.15. Orientation of a compressed glass plate in a beam of plane polarized light.

111

indices in this case (so that the two speeds are given by c/n_y and c/n_z, where c is speed of light in vacuum) then

$$n_z - n_y = R\sigma_y n$$

and R is a constant characteristic of the material. R is called the *stress-optical coefficient*. Some typical values for various materials are given in table 4.2.

Soda–lime–silica glass	2·5
Silica glass	3·4
Perspex	5
Epoxy Resin	7
Bakelite	53
Gelatin	2000–14 000

Units are micrometre path difference per metre of path length per MN m^{-2}, $R = (n_z - n_y)/n\sigma_y$.

Table 4.2.   Stress-optical coefficients.

Now suppose a beam of monochromatic light, which is plane-polarized with the vibration direction OP at some angle to the y axis, is passed through the loaded plate in the x-direction (fig. 4.16).   The plane-polarized light will be split into two components inside the birefringent plate, one with vibrations parallel to the y axis and the other with vibrations parallel to the z axis.   These components of the original beam will travel at different speeds, and will have different wavelengths, inside the plate ; when they emerge the vibrations in the y and z-directions will not be in phase so that the emergent beam of light is elliptically polarized.   If the beam then meets a second Polaroid sheet oriented so that only vibrations perpendicular to OP are transmitted, some light will nevertheless pass through, for the vibrations in the y and z-directions will have components perpendicular to OP (see plan in fig. 4.16 b).   Beyond the analysing Polaroid we have two waves with vibrations in the same direction, both obtained from a single beam of plane-polarized light ;   they will therefore interfere.   The difference in optical path travelled by the two components is $d(n_z - n_y)$, where d is the plate thickness, and the phase difference will be

$$\delta = \frac{2\pi}{\lambda} d(n_z - n_y) = \frac{2\pi d}{\lambda} R\sigma_y.$$

Since the two components are of equal amplitude ($= A \cos\theta \sin\theta$, see fig. 4.16 b), complete destructive interference (giving zero intensity) will occur beyond the analyser whenever d and σ_y are such that

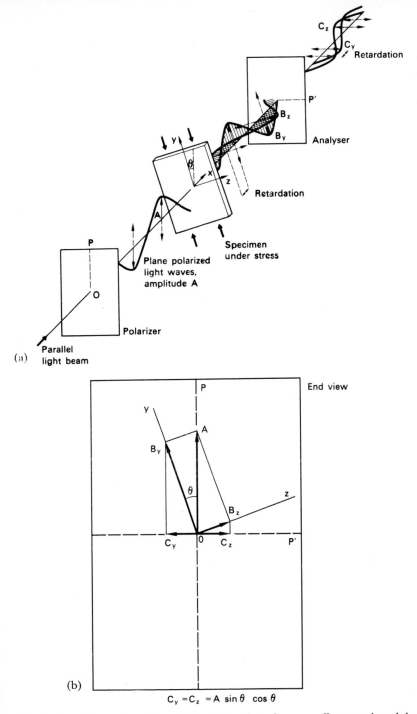

$$C_y = C_z = A \sin \theta \cos \theta$$

Fig. 4.16. Illustrating the origin of the interference effect produced by a stressed sample between crossed polarizers; (b) shows the projection of the wave vectors on to the plane perpendicular to the incident beam.

$\delta = 0$, 2π, 4π . . ., etc. and constructive interference (giving a maximum intensity) when $\delta = \pi$, 3π. . . .   In practice for a plate under uniform stress, δ will most likely have some intermediate value.   The magnitude of δ can be found by using a calibrated ' compensator ' in front of the analysing Polaroid to introduce an equal but opposite phase difference between the two components and therefore restore the overall δ to zero.   Compensators made from naturally birefringent crystals and having a continuously variable effective thickness are particularly useful.

If a beam of polarized white light is passed through a birefringent plate then the light emerging from the analyser is coloured : partial destructive interference will occur for a band of wavelengths around $\lambda_d = (d/p)(n_z - n_y)$, $p = \delta/2\pi = 1, 2, 3$ . . ., so that these components of the original white light will be partially destroyed.   The colour of the emergent light changes with the position of the destroyed band and becomes very hazy and ' washed-out ' for the higher orders (larger values of p) where the band becomes narrower.

When, for a stressed plate, the stress in the y-direction varies over the width or depth of the plate then, provided $dR\sigma_y$ is large enough, a series of fringes will be produced, fig. 4.17.   Each fringe is a contour line of constant phase difference.   With these more complex stress distributions, where there are additional tensile stresses in the z and x-directions or shearing stresses in the yz plane, the relation between the phase difference δ and the stress components is more involved but the fringe pattern produced is related to the magnitude and distribution of the stresses within the sample.

The stress-optical coefficient of glasses is relatively small (table 4.2) ; for a soda–lime–silica glass plate 10 mm thick, a uniaxial tensile stress of 20 MNm^{-2} would be required to reach the first-order dark fringe for $\lambda = 600$ nm.   It is essential to detect the presence of stresses much smaller than this in the course of glass-working operations.   The normal limit of allowable stress in ordinary glass-ware is $\sim 1/20$ of the breaking strength or about 2 MN m^{-2}.   For optical glass even lower limits are demanded in order to avoid the disturbing effects of birefringence.

Greater sensitivity is achieved by using white light and inserting between the polarizer and analyser a ' sensitive tint ' plate.   This is a thin plate of naturally birefringent material such as mica and the thickness and orientation with respect to the polarizer is chosen so that, in the absence of an additional birefringent specimen, first-order destructive interference occurs in the yellow part of the spectrum and the light emerging from the analyser is magenta.   When additional birefringence is introduced the ' missing ' band moves ; the colour of the emergent light changes towards blue or red and the eye is particularly sensitive to changes in hue in this part of the spectrum.

114

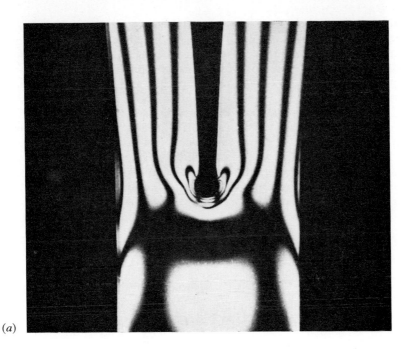

(a)

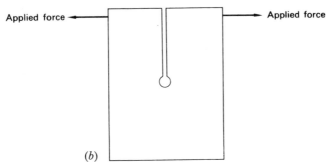

Applied force ←————————————————→ Applied force

(b)

Fig. 4.17. The photoelastic effect. (a) Interference fringes produced by a loaded sample (a double cantilever beam) of Perspex between crossed polarizers in monochromatic light ; (b) the shape of the sample and loading arrangement.

4.7. *Coloured glasses*

It is believed that a large number of different coloured glasses were in use centuries before the appearance of the clear transparent variety. Impurities commonly present in the raw materials will often result in a coloured glass and removal of these impurities or control of their effect on the colour requires some technical sophistication. In the manufacture of filter glasses for use in signals or in photography, precise control of the colour is essential. Knowledge of the compositions and melting conditions required to produce a particular colour has been built up by trial and error and in some cases the essential tricks required to produce a particular shade of colour are still regarded as 'trade secrets'.

Nevertheless, the basic physical mechanisms responsible for the selective absorption or selective scattering which produces the coloration are understood in principle. The practical difficulties and uncertainties are related to the identification and control of those factors which, in glasses, influence the exact position or width of the selective absorption band. Quite small changes in the position or width of an absorption band may profoundly effect the visual colour, fig. 4.18.

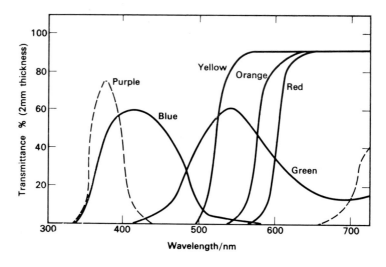

Fig. 4.18.   Absorption curves for some coloured glasses.

There are two fundamentally different ways of producing coloured glasses : the first involves the formation of a dispersion of fine particles within the glass and the second involves dissolving in the glass those ions which give rise to coloured aqueous solutions and coloured crystals. This second method is basically just the introduction of electrons which have energy level transitions (or on the wave picture, resonant frequencies) in the visible range, i.e. an energy gap between quantum levels of about 3 eV to 1·6 eV, or resonant frequencies $7·5 \times 10^7$ to 4×10^7 Hz which correspond to wavelengths from 400 to 720 nm.

116

For visible coloration by selective absorption, the absorption must occur over a range of frequencies ; a single, or even several *sharp* absorption lines however strong will not modify white light sufficiently to be detected by the eye.

Precipitation colours in glasses

In a dispersion of particles which are comparable in size to the wavelength of light, an incident beam of light suffers a combination of diffraction, diffuse reflection and scattering, and this varies in a complex way with the wavelength of the light and with the size and properties of the particles. Glasses containing a dispersion of colloidal particles exhibit particularly rich and beautiful colours.

A splendid red glass can be produced by adding gold (about 1 part in 5000 by mass) or a gold salt to the batch for a lead–silicate glass. The glass first produced by the normal melting and cooling process is colourless and the red colour is developed by reheating ; a process known as ' *striking* ' the colour. At high temperatures, the gold is distributed as isolated atoms or ions, i.e. is dissolved, in the glass. The solubility decreases as the temperature falls but during normal cooling there is insufficient time for aggregation of the gold atoms to occur and they are retained in solution. Reheating leads to the precipitation and growth of small particles of metallic gold which give the glass a strong absorption in the green, and hence a rich red colour. If the heating is continued, the particles increase in size and the colour changes to blue ; eventually the particles coagulate and, since gold in bulk has a maximum reflectivity in the yellow, the glass appears brown. The red colour of well-developed gold–ruby glasses is very intense and these glasses can be used as a coating on a base of colourless glass.

Because of the high cost of gold, other ruby glasses are now more common ; copper–ruby contains a colloidal dispersion of copper metal and cuprous oxide (Cu_2O), and so-called selenium–ruby contains particles of CdS in which small amounts of CdSe are dissolved. As the proportion of CdSe in the precipitated CdS crystals is increased, colours ranging from yellow to orange to deep ruby are produced. It is thought that the colour of these glasses is due to the colour of microscopic particles rather than to ' colloidal colouring ' as in a gold–ruby. Melting of the selenium–ruby glasses requires very careful control of the glass composition and the melting conditions to avoid, or to allow for, the loss of selenium by volatilization.

Photosensitive and photochromic glasses

The precipitation of colloidal metal crystals from glasses containing small amounts of copper, silver or gold in solution occurs more rapidly and at a lower temperature if the glasses are first irradiated with ultraviolet light. Small amounts of CeO_2 and Sb_2O_3 in the glass increase the sensitivity. The effect of u.v. irradiation is to form metal atoms which agglomerate to form a nucleus for the growth of small metal crystals during heat treatment. Photochemical reactions of the kind :

$$Cu^+ + h\nu \rightarrow Cu^{2+} + e$$
$$Cu^+ + e \rightarrow Cu$$

and

$$Cu^{3+} + Cu^+ + h\nu \rightarrow Cu^{4+} + Cu$$

occur in the glass.

If a plate of unstruck glass is selectively irradiated through a mask or a photographic negative, an image can be developed by subsequent heat treatment.

Photosensitive glasses are also produced in which the initial precipitation of metallic particles is followed by the growth of small silicate crystals from the glass itself, crystals such as lithium or barium disilicate ; these make the material opaque in the regions which were strongly irradiated. The discovery that for suitable initial glass compositions these opaque regions were much more rapidly dissolved in hydrofluoric acid has led to a useful technical process for ' chemically machining ' glass. Very intricate patterns in relief or perforation, e.g. accurate grooved scales or closely spaced holes, can be produced by selective exposure to ultra-violet light followed by heat treatment and then etching with dilute hydrofluoric acid.

Glasses have recently become available commercially which darken significantly on exposure to bright sunlight, but revert to a clear transparent form when the brightness falls. These *photochromic* glasses contain finely dispersed small crystals, such as silver bromide, which are light sensitive ; the light will reduce silver ions in these crystals to silver atoms by the same process that occurs in the silver halide crystals in a photographic emulsion. The darkening of the glass is due to the formation of a latent image of metallic silver but the process is reversible in a photochromic glass because the silver and bromine atoms formed by the photochemical reaction cannot readily diffuse apart and will react to form silver bromide again when the intensity of the radiation is reduced.

Both photosensitive and photochromic glasses have great potential in technical as well as commercial applications : photosensitive glasses enable very high resolution to be achieved in making scales and graticules and, if the relaxation time can be reduced, photochromic glasses offer special advantages as optical data-storage devices.

Solution colours in glasses

The quantized energy levels available to the electrons in an isolated atom fall into a series of distinguishable groups, called the K, L, M, N . . . *shells* ; the outer shells can be divided into subsidiary groups

118

or sub-shells. Although in general there are several quantum states corresponding to each level of energy, only one electron can be accommodated in each separate quantum state. Thus a series of different atoms which have progressively increasing atomic number, and therefore increasing numbers of extra-nuclear electrons, will have electron distributions which correspond to progressive filling of these energy level shells. The chemical and the optical properties of isolated atoms are closely related to the electronic configuration in the atoms, in particular to the number of electrons in the outermost, incompletely filled shell. The periodicity observed in the chemical properties of the elements when they are listed in ascending atomic number is related to the filling of these shells and the periodic recurrence of the same number of electrons in the outermost shell. It is the excitation of these outer electrons into higher normally empty energy levels and their decay back into lower, empty levels which is responsible for the characteristic atomic spectral lines in absorption and emission, respectively. Atoms of the alkali metals all have one electron in addition to the number required to form completely filled shells (or filled shells plus a major sub-shell). For example, Li has two electrons filling the K shell plus one in the L shell ; Na has two electrons filling the K shell, eight electrons filling the L shell plus one in the M shell. The characteristic series of spectral lines of Li and Na and of hydrogen follow the same kind of pattern because the electron transitions involved in all three kinds of atoms are those of the single outermost electron.

Ions in the first series of transition metals, viz. Sc, Ti, V, Cr, Mn, Fe, Co, Ni and Cu, frequently produce coloured solutions and crystals. Not only are many of the inorganic salts of these elements coloured, but their presence as an impurity in other, normally colourless, crystals can produce pronounced colouring. The gem stones, ruby and blue sapphire, are both aluminium oxide, Al_2O_3 : ruby owes its red colour to the presence of traces of chromium and blue sapphire to traces of titanium.

The elements of this series have an incompletely filled inner sub-shell of electron energy levels (the 3d orbitals), and the broad optical absorption bands responsible for the colour of their compounds are due to electron transitions between these levels. For an isolated atom or ion, such transitions are ' forbidden ' but for ions in solutions or in solids, where they are surrounded either by dipolar molecules or by other ions, the electric field from neighbouring charges changes the transition probabilities so that ' weak ' absorption can occur. The field from the surrounding charges also affects the energies of the levels : fluctuations in the position of neighbouring ions in a solid, due to thermal vibrations, causes fluctuations in the energy gap between levels, so that the size of the gap varies from one ion to another at any particular instant, or for any one ion over a period of time large compared with

119

the period of the thermal vibrations, i.e. $\gg 10^{-13}$ s. Since the absorption of a photon, due to excitation of an electron across the gap between these levels, occurs in a time $\ll 10^{-13}$ s, the collection of ions as a whole will show a broad band of absorption and this leads to richly coloured solutions or solids.

Ions of the lanthanide and actinide series also have an incomplete sub-shell, and when these ions are present in crystals or in glasses, transitions between the levels in the sub-shell give rise to an optical absorption band. However, for these ions, the incomplete sub-shell is screened by an outer, filled sub-shell ; the energy levels of electrons in the incomplete sub-shell are therefore not greatly altered by the positions of surrounding ions and the absorption peaks are much sharper. A very high quality optical glass containing neodymium ions has been produced recently as an alternative to synthetic ruby rods for use in lasers.

Transition metal ions in solution or in solids may have a broad absorption band in the optical region but the position of this band depends on both the strength and the symmetry of the field due to surrounding dipoles or ions. Therefore the colour produced by a given ion depends on the nature and the arrangement of surrounding ions. For example, the divalent cobalt ion produces a pink colour in crystals when it occupies a site surrounded by six oxygen ions distributed at the corners of an octahedron, and a blue colour when surrounded by four oxygen ions at the corners of a tetrahedron. Similar changes occur for cobalt ions in glasses ; in a borosilicate glass, cobalt gives a pink colour due to the predominance of CoO_6 groups, while in a potash–silica glass it gives a purple colour because both CoO_4 and CoO_6 groups are present. The addition of small quantities of copper to a soda–lime–silica glass produces blue colours similar to that of hydrated copper sulphate crystals, $(CuSO_4 . 5H_2O)$. But the same amount of copper in a borosilicate glass produces green colours rather like the green of the copper mineral, malachite, $(Cu(OH)_2 . CuCO_3)$.

Many of the transition metals change their oxidation state (valency) easily in glasses. Mn, Cr and Cu are all used commercially to produce coloured glasses and, depending on their valency state, they can exist as part of the network or as network modifiers. In soda–lime–silica glasses, Mn^{3+} gives a deep purple and Mn^{2+} a faint pink ; Cr^{3+} gives green and Cr^{6+} yellow ; Cu^+ is colourless and Cu^{2+} gives blue.

The relative proportions of ions in different oxidation states and/or the coordination of the colouring ions can be controlled by varying the glass composition and the melting conditions. Herein lies the technical sophistication (or the art !) of making coloured glasses of accurately reproducible hue. In many cases, the craftsman is still ahead of the scientist and the ion or ion group responsible for a particular colour is still uncertain. Some of the traditional colouring agents are listed in table 4.3.

Colour	Colouring agent
Opal	Precipitated calcium or sodium fluoride *or* ' droplets ' of phosphate glass
Red	Colloidal gold, colloidal Cu and CuO_2 or microscopic crystals of CdS with CdSe
Yellow	Microscopic crystals of CdS
Green	Chromium or copper ions
Blue	Cobalt or copper ions
Violet	Manganese ions
Amber (e.g. brown beer bottles)	Complex ion group formed by adding iron and sulphur and carbon (as reducing agent) to glass batch
Black	High concentrations of manganese and copper, iron or cobalt
Fluorescent (e.g. for the tubing in fluorescent lights)	Uranium

Table 4.3.

Colourless glass

Iron is present in most sands as an impurity and the green/blue tint detectable in thick blocks of ' ordinary ' glass, or in plate glass viewed end-on, is due to the presence of ferrous ions. The Fe^{2+} ion has a very strong absorption band in the infra-red and the tail of this runs into the red end of the visible region, producing the faint green coloration. For ' colourless ' glass the iron content of the sand must be less than ~ 0.04 per cent Fe_2O_3 and special care must be taken to avoid contaminating the batch with particles of iron or iron oxides from the mixing and handling machines, or from the furnace. The effect of residual iron content can be reduced by melting the glass under oxidizing conditions and adding As or Sb in order to maximize the proportion of Fe^{3+} ions, which gives a less intense yellow colour. Alternatively the faint green colour can be masked by the addition of Se and Co so as to produce the complementary colour—pink. For optical quality glass, which must have a very low, uniform absorption coefficient throughout the visible spectrum, it is essential to use sands with even lower concentrations of iron.

4.8. *High energy radiations and colouring of glasses and crystals*

It has long been known that normally clear, ' white ' glasses can become coloured on prolonged exposure to very strong sunlight. Michael Faraday noted the development of a purple colour in a glass containing manganese and correctly ascribed the change to the conversion of Mn^{2+} into Mn^{3+} ions in the glass by the ultra-violet component of sunlight. More recently, ' discoloration ' due to the

121

irradiation of glasses by X-rays and γ-rays has been studied in some detail. The effect can impair the usefulness of glass windows in nuclear equipment, but has been used to advantage in the development of special glasses for use as dose-meters.

Two principal effects are responsible for the development of colour during irradiation : electrons displaced by the high-energy radiation may either change the oxidation state of a transition metal ion, or they may become trapped at special sites within the glassy network. Similar colouring effects occur in crystals of the alkali halides and the concepts being used to explain the observations in glasses were developed first for these crystals.

Sodium chloride is normally colourless but after irradiation with X-rays or γ-rays it becomes, temporarily, yellow ; potassium chloride becomes blue. Similar, but more intense and permanent, colours can be produced in these crystals by heating them in an atmosphere of the alkali metal vapour. The crystals then become slightly non-stoichiometric, with more Na^+ or K^+ than Cl^- ions in the structure. The change in composition is accommodated by the formation of an excess of vacant Cl^- lattice sites and by the addition of electrons to preserve the electrical neutrality of the whole crystal. These electrons become ' trapped ' in orbits around the vacant Cl^- sites and form so-called *F-centres*. The anion vacancy is like a localized positive charge in the lattice and an electron is attracted into an orbit around this site in much the same way as is the electron around the proton in a hydrogen atom (fig. 4.19). There is a series of hydrogen-like orbits available to the electron around the vacancy and a trapped electron can be excited to a higher energy by the absorption of a photon of appropriate energy.

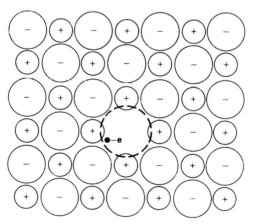

Fig. 4.19.   Illustration of an F-centre : an electron trapped in an orbit around a negative-ion vacancy.

122

Absorption due to F-centres in alkali halides occurs over a relatively wide band of wavelengths ($\Delta\lambda \sim 100$ nm) in the visible range, because the energies of the discrete quantum states of the electron (analogous to the ' stationary states ' of the electron in the Bohr theory of the hydrogen atom), and therefore the energy differences between the states, depend upon the spacing of the positive ions surrounding the vacant Cl^- site ; the positive ions undergo thermal vibrations but the excitation of the electron occurs in a time which is very much shorter than the period of vibration of the ions, so that the excitation energy varies with time as the instantaneous spacing of the surrounding positive ions changes. This broadening of the absorption line is analogous to that discussed earlier for the transition metal ions in a solid.

The colour of crystals which have excess alkali metal ions is due to F-centres but other more complicated types of *colour centre* can also exist in alkali halide crystals. Under X- or γ-irradiation electrons may be knocked off a halogen ion and become trapped at a halogen ion vacancy forming an F-centre. The halogen ion from which the electron came may transfer its loss to an adjacent ion : we can regard the ' missing electron ' as though it were a real entity with a positive charge. It is called a *positive hole*. A positive hole can be trapped at an alkali ion vacancy, forming a V-centre. In NaCl and KCl, V-centres produce absorption bands in the ultra-violet. Irradiation produces F- and V-centres but the colour is produced by the absorption of the F-centres and is only temporary because the electrons and positive holes eventually recombine ; optical and thermal excitation will occasionally release charges from their traps and the colour will bleach.

Analogous colour centres are formed in glasses during irradiation with X- or γ-rays ; a vacant alkali ion site may trap a positive hole and a vacant oxygen ion site may trap an electron. However, the process of identifying the types of colour centre responsible for each of the observed absorption bands is much less advanced.

CHAPTER 5
the elastic properties of glass

5.1. *Introduction*

ITS mechanical properties determine to a large extent the use that is made of any given material.   If we were selecting a material with which to build a bridge or even a set of bookshelves we should obviously pay particular attention to its behaviour under load.   We should want to be sure that the material in a suitable size and shape was both strong enough to support the expected loads and stiff enough to prevent any large elastic deflections.   It might also make for easier and therefore cheaper construction if we selected a material which could be sawn and drilled.   Additional properties such as resistance to chemical attack by the atmosphere, appearance, cost, etc., although important would not restrict our choice to the same extent as the mechanical behaviour.

Although perhaps not quite so obvious, it is also true that the mechanical properties are often a major factor in the selection of a material for many applications which are not primarily structural. Glass is usually preferred to transparent plastics for most windows and lenses because of its much greater resistance to scratching ; on the other hand, the transparent parts of safety goggles and riot-shields are made of Perspex or other transparent plastic which are less prone than glass to shatter under impact.

As these examples illustrate, the mechanical properties of significance in practice are often much more complex than those basic and easily defined ones like the elastic constants and the ultimate tensile strength. Various empirical tests have been devised to enable engineers to compare those characteristics of different materials which have been found important in practice.   There are hardness tests to assess the scratch resistance or the penetration resistance of a material, impact tests to compare the total energy expended in breaking different materials, fatigue tests to provide a measure of the lifetime of materials under an oscillating load, and so on.

While these tests are invaluable to the engineer and the designer, few as yet can be interpreted in a fundamental way.   In this chapter and the following one we shall be concerned chiefly with the simple basic mechanical properties :   the elastic behaviour and the strength and fracture of glass.

124

5.2. *Elastic behaviour of solids*

All materials deform, i.e. change their shape or size or both, under externally applied forces. Solids will recover their original shape and size when the forces are removed, no matter how they were distributed, provided the deformation was not too large. In fact we might define a solid as that class of material which exhibits a recoverable deformation when subjected to a small shearing force. Fluids under shear, however small, deform continuously with time for as long as the force is applied and do not recover their original shape when the force is removed ; they have no rigidity.

The deformation of a solid by an applied force is said to be elastic if it is completely reversible. When the fractional change in dimensions (the *strain*) is small, then for nearly all solids it is proportional to the applied force per unit area (the *stress*) : this is Hooke's law. The ratio stress/strain is the elastic modulus and, for isotropic solids, only two independent moduli are required in order to calculate the elastic deformation resulting from any arbitrary distribution of applied forces. In practice however, four elastic constants (Young's modulus, the rigidity or shear modulus, the bulk modulus and Poisson's ratio (fig. 5.1)) are in common use and are convenient for dealing with many practical problems. They are related by the following equations :

$$Y = 2G(1 + \nu)$$

and

$$Y = 3K(1 - 2\nu).$$

Hydrostatic compression or extension of a solid corresponds to the displacement of all the atoms or ions from the normal minimum energy position indicated in fig. 2.6. As we have already seen in the discussion of thermal vibrations the net force on an atom can be written (equation (2.12)) as

$$F = -\beta x + \gamma x^2,$$

where x is the increase in interatomic spacing. The strain is simply x/d_0, where d_0 is the normal equilibrium separation, so that for small strains x is small, γx^2 is negligible and hence F is proportional to x. Thus for small strains the ratio stress/strain will be constant but as the strain and therefore x increases, we can expect the ratio to decrease if x is positive and to increase if x is negative. Thus at high strains the elastic modulus should increase in compression and decrease in tension. The values of β and γ indicate that these changes should be significant above an elastic strain of about $\frac{1}{2}$ per cent.

Not all solids have a structure such that deformation by an applied force is opposed from the start by primary bonds : in some materials deformation can occur through a rearrangement of the atoms which does not initially change their separation. Rubber, for example, consists of

125

tangled coils of long-chain molecules ; atoms within any one molecule are strongly bound but there is only relatively weak dipolar bonding between the molecules. Rubber has a very low Young's modulus and also exhibits elastic strains up to several hundred per cent. The extension in the direction of an applied tensile stress is due to the

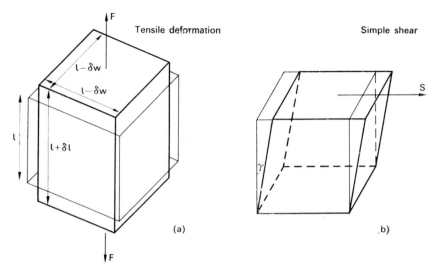

Fig. 5.1.

(In each figure the faint lines show the unstressed body which is a cube of side l.)

(a) *Tensile deformation*
  Longitudinal (uniaxial) force $= F$.
  Longitudinal strain $= \delta l/l$.
  Lateral strain $= \delta \omega/l$.

$$Y = \text{Young's modulus} = \frac{\text{longitudinal stress}}{\text{longitudinal strain}}$$

$$= \frac{F}{\text{area}} \div \frac{\delta l}{l} = \frac{F}{l\delta l}.$$

$$\nu = \text{Poisson's ratio} = \frac{\text{lateral strain}}{\text{longitudinal strain}} = \frac{\delta \omega}{\delta l}.$$

(b) *Simple shear*
  Bottom face of the cube is fixed and a tangential force S applied to the top face.
  Shearing stress $= S$.
  Shear strain $= \tan \gamma \approx \gamma$.

$$G = \text{shear (rigidity) modulus} = \frac{\text{shear stress}}{\text{shear strain}}$$

$$= \frac{S}{\text{area}} \frac{1}{\gamma} = \frac{S}{l^2} \frac{1}{\gamma}.$$

126

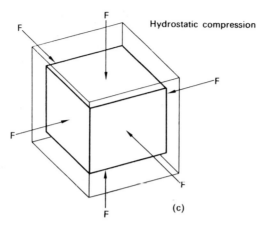

Fig. 5.1. (*cont.*)

(c) *Hydrostatic compression*
Equal forces on all faces.

$$K = \text{Bulk modulus} = \frac{\text{hydrostatic pressure}}{\text{volume strain}} = \frac{F}{\text{area}} \frac{v}{\delta v}.$$

straightening of the molecular chains. Eventually, at very large strains, the chains become aligned in the direction of the applied tension and thereafter further extension is opposed by the strong forces between the atoms within each chain and the material becomes very much stiffer. Other materials such as wood, foamed plastics or foamed glass have a cellular structure and these materials may deform as the cell shape alters due to the elastic bending of the cell walls under an applied stress.

Both long-chain polymers and cellular materials may suffer large strains under comparatively low stresses but their elastic moduli in tension and compression are usually quite different. However, for 'atomically compact' materials such as metals, ionic crystals and glasses we can expect that elastic deformation will involve changing the interatomic spacings and therefore stretching or compressing the primary interatomic bonds. The elastic moduli at low strains will be the same in compression as in extension and will be proportional to the slope of the force–displacement curve or to the curvature of the energy–displacement curve like those in fig. 2.5. In principle we expect these materials to depart from linear elastic behaviour at strains of the order of $\frac{1}{2}$ per cent but in practice, for normal samples, fracture or irreversible deformation occurs long before strains of this magnitude are reached.

Even at low strains most solids depart very slightly from ideal elastic behaviour in another way : the strain is observed to change with time under constant stress. This is a *delayed elastic* deformation ; it is

127

recovered when the stress is removed although the return to zero strain also takes place over a period of time. The variation of elastic strain with time, called anelasticity, is analogous to the variation of dielectric polarization with time which we discussed in Chapter 3, and a similar phenomenological description can be used for both these effects. Under an alternating applied stress, delayed elastic relaxation will lead to an absorption of energy ; the free oscillations of a solid, for example in the form of cantilever beam or a torsion suspension, will decay with time due to this absorption of energy even when external damping from air resistance, etc., is negligible. The energy absorption effect is sometimes referred to as *internal friction*. We shall discuss some of the mechanisms responsible for anelastic effects in section 5.4 when dealing with the anelasticity of glasses. In glasses below the transformation range of temperatures, the delayed elastic strain is a very small fraction of the total elastic strain and we shall ignore it in the following section.

5.3. *Elastic behaviour of glass*

At temperatures well below their softening points and for low strains, glasses obey Hooke's Law ; the ratio stress/strain is a constant and there is no observable permanent (plastic) deformation of a macroscopic sample up to the stress at which fracture occurs. For normal samples of commercial glasses the fracture stress is about $50 \, MN \, m^{-2}$, which corresponds to an elastic strain of about $0 \cdot 1$ per cent. As we shall see later, specially prepared specimens can support very much higher stresses, approaching $10^4 \, MN \, m^{-2}$, and significant departures from Hooke's law become apparent. There is, however, still no detectable plastic flow, so that although the strain is not exactly proportional to the stress, it is still elastic : a specimen recovers its original shape and size when the stress is removed.

Some values of the elastic constants for different types of glass are given in table 5.1. There is very little change with composition, although from the earlier arguments we should have expected that silica would be significantly stiffer, in view of the strong Si–O bonds, than

Glass	Young's modulus $10^{-10} Y/N \, m^{-2}$	Rigidity modulus $10^{-10} G/N \, m^{-2}$	Poisson's ratio ν
Silica	7·4	3·2	0·16
Aluminosilicate	9·1	—	0·26
Borosilicate (Pyrex)	6·1	2·5	0·22
Soda–lime–silica	7·4	3·1	0·21
Lead–silicate	6·1	—	0·21

Table 5.1.   Elastic constants of glasses.

128

the modified glasses. The value of Poisson's ratio for silica is also remarkably low. It has been suggested that for very open networks, elastic deformation may involve some pivoting of the tetrahedra which form the walls around the larger holes. This would give a deformation process analogous to the buckling of the walls in a cellular material, but on a much smaller scale, and would produce a strain additional to that resulting from the change in interionic spacing.

Some evidence in support of this hypothesis is to be found in the behaviour of silica glass at high strains. Under high hydrostatic pressure the bulk modulus decreases with increasing pressure and at very high pressures, $\sim 10^{10}$ Pa, a permanent increase in density occurs :

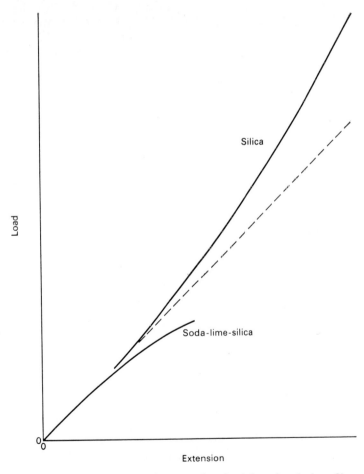

Fig. 5.2. Load versus elastic extension for ' flaw-free ' glass fibres.

129

the open network structure apparently collapses since the average Si–O distance is the same in the compacted material, after removal of the pressure, as in the original glass. Young's modulus for silica also shows anomalous behaviour at high strains. This is illustrated in fig. 5.2 which is based on some results reported recently by F. P. Mallinder and B. A. Proctor of Rolls-Royce Ltd. Research Laboratories. Silica and soda–lime–silica glass fibres were specially prepared and were stressed at − 196°C (the boiling point of liquid nitrogen) in order to achieve the highest possible strains before fracture. Strains of ∼ 12 per cent for the silica fibres and ∼ 6 per cent for the soda–lime–silica fibres were achieved. The fibres were not of constant cross-section so that calculation of numerical values for Young's modulus, from the measured loads and extensions, is not straightforward. However, it is clear from the curvatures in fig. 5.2 that at high strains the soda–lime–silica fibres become less stiff while the silica ones become stiffer. A decrease in stiffness at high strains is in keeping with the expected variation in interatomic forces. Very fine metal ' whiskers ' have been produced which are much stronger than ordinary specimens of the same metal and they will deform elastically to quite high strains. Figure 5.3 is

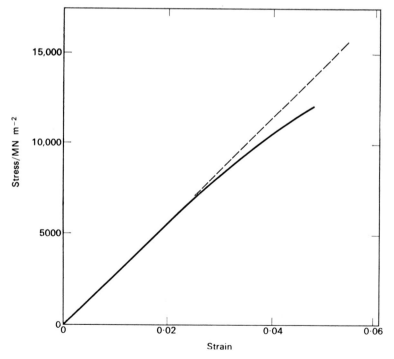

Fig. 5.3.   Elastic stress–strain curve for an iron whisker.

130

the stress–strain curve for an iron whisker which also exhibits a decrease in stiffness at high strains.

The anomalous increase in modulus observed with silica fibres is consistent with the hypothesis that, at low strains, the elastic deformation is augmented by distortion of the rings of SiO_4 tetrahedra. Such distortion will cease when the oxygen ions from opposite sides of a ring come into contact and the structure will therefore become progressively stiffer, as the distortion of successively larger rings reaches saturation, until bond-stretching becomes the only mechanism available. In soda–lime–silica glasses distortion of rings of tetrahedra will be much less common because many of the rings will be 'filled' by one or more relatively large modifier ions.

5.4. *Anelasticity*

Delayed elastic effect

In common with many solids, glasses show a small delayed elastic effect : a rapidly applied tensile or shearing stress produces an initial, virtually instantaneous, elastic strain which then increases slowly with time and eventually reaches a stable value. When the applied stress is removed, the initial rapid decrease in strain is followed by a further slow relaxation, fig. 5.4. Alternatively, when a specimen is elastically deformed and then held at constant strain, the stress required to maintain the strain falls with time. Delayed elastic effects can be studied either by following the delayed strain at constant stress or the relaxation of stress at constant strain. Since a specimen recovers its original dimensions in time after the removal of an applied stress, the behaviour can be represented by elastic constants which are functions of time.

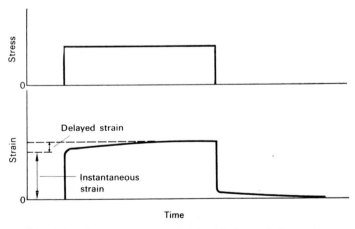

Fig. 5.4.   Instantaneous and delayed elastic deformation.

131

We can define an *unrelaxed modulus*, M_I, as the ratio—(constant stress/initial strain) or (initial stress/constant strain) and a *relaxed modulus*, M_R, as—(constant stress/final strain) or (final stress/constant strain). For most inorganic materials the delayed strain is very

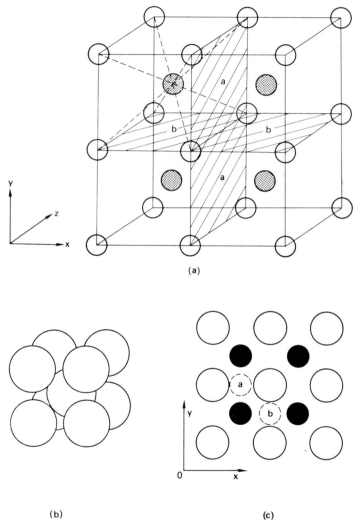

(a)

(b)   (c)

Fig. 5.5. Illustrating the distortion of a crystal by an interstitial atom. (a) and (b) show the atomic arrangement in iron crystals ; (c) is a two-dimensional analogue of an ionic crystal. The structures in (a) and (c) elongate in the x-direction when interstitial sites a are occupied and in the y-direction when sites b are occupied.

132

small, of the order of 1 per cent of the initial elastic strain so that $(M_I - M_R)/M_R \lesssim 0.01$. The departure from a simple time-independent form of Hooke's law is therefore not easy to detect and for many practical purposes it can be ignored. Very much larger delayed elastic effects are observed with long-chain polymers where elastic strain is produced by the uncoiling of the molecular chains ; this process requires an activation energy and quite slow uncoiling will occur in some temperature range for any given polymer.

In simple crystalline materials, various processes involving lattice defects contribute to the anelasticity. A particular example may help to show how this can occur. Figure 5.5 a illustrates the arrangement of atoms in a crystal of iron ; the circles represent the positions but not the actual sizes of the atoms. In the real crystal the iron atom in the middle of each small cube actually touches those at the corners of the cube but the corner ones do not touch one another, as in fig. 5.5 b. At normal temperatures, impurity atoms of carbon or nitrogen, which are often present in iron, are located in interstitial sites such as a or b. If the interstitial atoms are in sites like a then the crystal is elongated in the x-direction ; if they are in sites like b the crystal is elongated in the y-direction. A simple two-dimensional analogy is shown in fig. 5.5 c ; again the interstitials in sites like a make the ' crystal ' longer in the x-direction. When a tensile stress is applied in the y-direction, the interatomic spacing parallel to the y axis becomes greater and, because of the Poisson contraction, that in the x-direction (and in the z for the three-dimensional case in fig. 5.5 a) becomes smaller. Interstitial atoms in sites a will then have a higher energy than those in sites b, because they will be squeezed harder by their nearest neighbours. Hence if atom-jumping can occur, an initially random distribution of interstitials will eventually change so that the majority are in the lower energy sites like b. This change in distribution of the interstitial atoms will elongate the crystal in the y-direction, that is an additional strain will appear in the course of time in the direction of the applied stress. When the stress is removed, the energy difference between the two types of site disappears and in time the interstitial atoms become randomly distributed again, so that the additional strain is removed.

A very similar mechanism contributes to the anelasticity of glasses. Although there is no regularity in the type of site occupied by the alkali metal ions, nevertheless when the network is under a tensile or shear stress, distortion of the irregular holes in the network will increase the energy of the modifier ions in some sites and decrease it in others ; as the modifier ions jump into the lower energy sites, changes in the shape of the surrounding ' rings ' of silica tetrahedra occur which will be such as to increase very slightly the strain in the direction of the applied stress.

There are relatively few instances where the small anelastic effects which occur in simple crystalline materials and silica-based glasses

K   133

affect the use of these materials ; occasionally very low internal friction is required, e.g. for bells, and these are usually made from special tin–copper alloys which satisfy this requirement. The main scientific interest in the anelasticity of simple inorganic materials arises because the effects provide a method of studying the mobility of some of the lattice defects or ions in these materials. Small but measurable anelastic

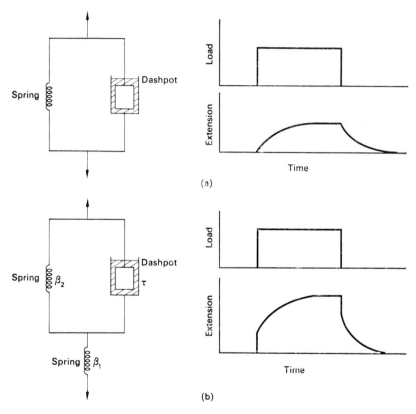

Fig. 5.6. Spring and dashpot models and the corresponding displacement–time curves.

effects can be produced by fairly small concentrations of mobile carbon atoms in iron even at temperatures where the jump frequency is only of order one per second ; in order to detect the movement of atoms with this jump frequency by a conventional diffusion experiment, in which changes in a concentration gradient through the solid are measured, a diffusion time of order 10^{10} s ≈ 300 years would be required !

134

Phenomenological theory
Delayed elasticity and the associated energy loss under oscillating stress can be represented by a phenomenological theory which is exactly analogous to that used in Chapter 3 for delayed dielectric polarization.   In fact, for the mechanical phenomena, we can envisage a simple physical system consisting of ideally behaved components which would exhibit the same kind of effects as real solids.   The components are ideally elastic springs and a cylinder and piston filled with a viscous liquid, called a dashpot.

Figure 5.6 shows the response to an applied load to be expected for two combinations of springs and dashpots.   If we put a spring in parallel with a dashpot, fig. 5.6 *a*, and apply a load, the displacement will increase with time until the extension of the spring is sufficient to give a restoring force equal to the applied force.   A combination of two springs and a dashpot as in fig. 5.6 *b* would produce an instantaneous and a delayed displacement when a load is applied or removed.   Movement of the piston of a dashpot involves shearing the liquid so that the higher the viscosity of the liquid the longer it takes for the displacement to reach a stationary value.   Equations representing the displacement of the model systems in fig. 5.6 are easily obtained.   For an ideal Newtonian liquid, the rate of shear is equal to (*shearing stress*)/(*coefficient of viscosity*), and the shearing stress on the liquid in a dashpot is proportional to the force on the piston.   The force on the piston of the dashpot in fig. 5.6 *a* is just the difference between the externally applied force and the restoring force due to the extension of the spring.   The force on the piston is zero when the displacement reaches saturation value, y_s, for any given applied force ;   for any other displacement, y_2, it will be proportional to $(y_s - y_2)$ since the spring is ideally elastic and therefore the restoring force will be proportional to the extension. Thus we have

rate of change of displacement $\propto$ force on piston $\propto (y_s - y_2)$

or

$$\frac{dy_2}{dt} = (y_s - y_2)/\tau.$$ (5.1)

When the load is removed $y_s = 0$, and

$$\frac{dy_2}{dt} = -y_2/\tau.$$ (5.2)

These equations are of exactly the same form as equations (3.25) and (3.26) respectively for the growth and decay of delayed dielectric polarization.   The solutions are

under a constant applied load $y_s = $ constant
and $y_2 = y_s[1 - \exp{(-t/\tau)}]$ (5.3)

135

and after removal of an applied load

$$y_2 = y_0 \exp\left(-t/\tau\right).$$ (5.4)

The constant τ is again a relaxation time and in this case it increases with the viscosity of the liquid in the dashpot.

For the combination in fig. 5.6 b the total displacement under a constant load will be

$$y = y_1 + y_2 = y_1 + y_s[1 - \exp\left(-t/\tau\right)],$$ (5.5)

where y_1 is the rapid displacement ($= F/\beta_1$) due to the upper spring, y_2 is the delayed displacement due to the spring and dashpot in parallel and y_s, the saturation value of y_2, will be F/β_2; β_1 and β_2 are the two spring constants. The unrelaxed 'elastic constant' for the whole system (in this case we use force/displacement since we have not defined the cross-sectional area or the unstrained length of the system) will be

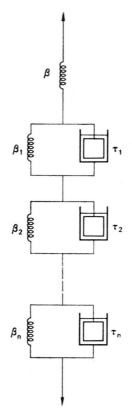

Fig. 5.7. Spring and dashpot model for the behaviour of a real material.

136

β_1 and the relaxed ' elastic constant ' will be $\beta_1\beta_2/(\beta_1+\beta_2)\approx\beta_1(1-\beta_1/\beta_2)$ if $y_1\gg y_2$ so that $\beta_2\gg\beta_1$.

Under an oscillating force, the displacement in the spring and dashpot model in fig. 5.6 b may lag behind the force and this will lead to an absorption of energy. Here, just as for the lag of dielectric polarization behind the electric field, the energy loss can be represented by a dissipation factor, tan δ, where δ is the phase angle of lag. Tan δ is again proportional to $\omega\tau/(1+\omega^2\tau^2)$.

We have outlined above the derivation of equations which represent the displacement of the spring and dashpot models as a function of time under constant applied force ; a complete set of equations can also be derived which represent the relaxation of the force at constant displacement. These equations form the basis of a phenomenological theory of anelastic effects in solids. In real solids there are invariably a number of different processes contributing to the delayed elasticity : the behaviour of a real solid is more like that of the spring and dashpot model in fig. 5.7 and may be represented by using a series of relaxation times rather than a single one.

Experimental methods

For many solids, including glasses, the delayed elastic strain is very small ; the total elastic strain which can be produced before fracture or other irreversible change occurs is often $\lesssim 0.1$ per cent and the delayed strain is typically ~ 1 per cent of the total. The maximum

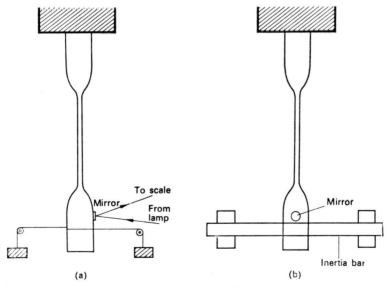

Fig. 5.8. Methods for measuring anelastic effects.

137

change in dimensions due to the delayed part of the elastic deformation is therefore only of the order of 1 part in 10^5; for a glass rod 1 m long the probable maximum increase in length before fracture occurs is about 1 mm, and the maximum delayed change is ~ 0.01 mm. Measurement of the change in length with time would require a sensitivity of the order of 1 part in 10^7 and detection of changes in length of the order of 0.1 μm even for a specimen 1 m long; this is not easily achieved !

One simple method of studying these small anelastic strains, which has been used by many research workers for all sorts of materials, is to stress a long wire or fibre in torsion rather than tension, fig. 5.8 a. The strain in the sample is then measured as the angle of twist; this angle can be measured very accurately using an optical level and a relatively small torsional strain in a long wire produces a large angular

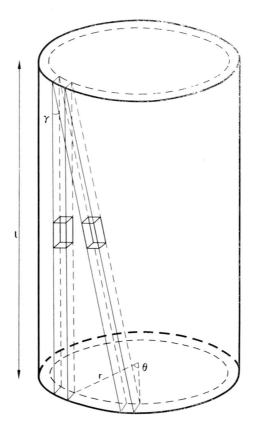

Fig. 5.9.   Torsion of a rod : each volume element suffers a simple shear strain given by $\tan \gamma \approx \gamma = r\theta/l$.

displacement of one end with respect to the other. Each volume element in a rod twisted about its axis, fig. 5.9, suffers a simple shearing distortion and the magnitude of the shear strain is proportional to the distance of the element from the axis of the rod ; the strain is zero on the axis of the rod and reaches a maximum of $r\theta/l$ at the surface, where r is the radius and l the length of the rod and θ is the angle of twist. However, if the maximum attainable elastic strain is ~ 0.001 so that $\theta = 10^{-3}$ (l/r) and the delayed strain is ~ 1 per cent of the total, then the change in angle of twist due to the delayed part of the strain will be 10^{-5} (l/r). Using an optical lever of length 1 m to measure angular displacements and for $l \sim 1$ m, $r \sim 0.1$ mm, the change in scale reading due to the delayed twisting will be $\sim 10^{-1}$ m. With this method it is, therefore, quite easy to follow the changes in the delayed strain using very simple apparatus.

If the relaxation times for the delayed processes in a given solid are already known, then the strain as a function of time at constant stress can be calculated using an equation similar to equation (5.5) but with a series of terms containing the different relaxation times. The reverse process, calculating the spectrum of relaxation times from observed changes in strain, is much more difficult. In principle measurement of the energy losses (the internal friction) as a function of frequency should provide a much more direct way of determining the relaxation times ; if there is a series of different mechanisms each with a distinct relaxation time at a fixed temperature, then the internal friction will rise to a peak whenever the frequency of the oscillating stress coincides with the reciprocal of one of the relaxation times. Unfortunately it is not at all easy in practice to apply alternating stresses which have a frequency that can be varied over a wide range. It would certainly involve using several very different techniques : measurement of the energy losses at high frequencies would correspond to determining the absorption of sound or ultra-sonic waves and at very low frequencies to finding the areas of stress–strain hysteresis loops. A convenient and commonly used alternative procedure is to use a fixed frequency and vary the temperature ; since most of the relaxation processes require an activation energy and for these $\tau \propto \exp(\phi/kT)$, we can characterize each process by the temperature at which it occurs for a fixed frequency. Or, by using a few different discrete frequencies, the activation energy, ϕ, can be found from the change in characteristic temperature with frequency. The simplest experimental procedure is to excite free oscillations in a specimen and measure the damping of the oscillations, due to internal friction, as a function of temperature. A fibre or wire specimen which supports an inertia bar or disc, fig. 5.8 b, can be set into torsional oscillation and the moment of inertia of the bar or disc can be adjusted so as to bring the natural frequency of oscillation to a convenient value. The free oscillations are damped : the amplitude

139

dies away because energy is absorbed by the specimen in each cycle. When the internal friction in the specimen is very low, the specimen and inertia member must be placed in vacuum in order that the damping due to the viscous drag of air does not dominate, but for many metals and for glasses other than pure silica the air damping is negligible compared with the internal friction.

Provided the damping is not too large and is independent of the amplitude of the free oscillations, the amplitude dies away exponentially with time and the angular deflection is given by an equation of the form $\theta = \theta_0[\exp{(-\gamma t/T)}]\cos{(2\pi t/T)}$, where T is the period (see fig.5.10). The *logarithmic decrement*, γ, is usually taken as the measure of the internal friction and

$$\gamma = \ln{(\theta_1/\theta_2)} = \frac{1}{n}\ln{(\theta_1/\theta_n)},$$

where θ_1 and θ_2 are the maximum angular displacements in successive cycles and θ_n is the angular displacement in the nth cycle. It can be shown that

$$\gamma = \pi\tan{\delta}.$$

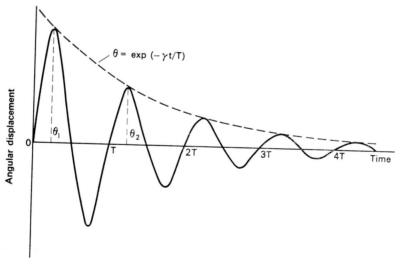

Fig. 5.10. Damped oscillations.

5.5. *Anelasticity in glasses*

The delayed elastic strains and internal friction are very small in glasses below the transformation region. In general the magnitude of the anelastic effects increases with the number of alkali metal ions present in the glass, being smallest for 'pure' silica and biggest for

140

' mixed-alkali ' glasses, and also increases as the temperature approaches the transformation range ; well-annealed samples show smaller effects than chilled ones of the same glass. Near the transformation range, irreversible viscous relaxation occurs in addition to larger, and faster, delayed elastic strain, fig. 5.11.

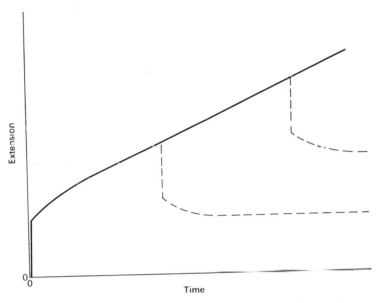

Fig. 5.11. Combination of delayed elastic relaxation and viscous flow under applied stress observed in the transformation range. The dotted curves show the change which would occur following removal of the applied load at two different times.

Measurements of the logarithmic decrement for free oscillations of fibres as a function of temperature show, for many glasses, two clearly defined broad absorption peaks superposed on a general background absorption which increases rapidly at high temperatures, fig. 5.12. Systematic investigation of the effect of composition has shown that the low temperature peak is associated with the jumping of alkali metal ions and the high temperature peak with non-bridging oxygen ions ; in simple binary glasses, SiO_2 + modifier oxide, both peaks increase in height in proportion to the concentration of the modifying oxide. The mechanism responsible for the background absorption at low temperatures has not been identified. The changes in the position of the internal friction peaks with the frequency of oscillation enable the activation energy associated with each relaxation process to be calculated.

141

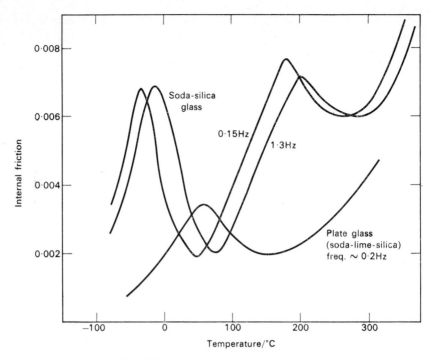

Fig. 5.12. Internal friction peaks in glass.

We have now seen that the jumping of alkali ions in glasses can be detected by studying the d.c. conductivity, the dielectric losses or the internal friction as well as by conventional diffusion techniques. The activation energies for Na$^+$ ion jumping in a silica network found from measurements of these different phenomena are all of the same order, ~ 1 eV per ion. There are no complete sets of all four measurements for samples of exactly the same glass but in some recent experiments with alkali–silica binary glasses, I. Mohyuddin and R. W. Douglas have shown that there is a small systematic difference, ~ 20 per cent, between the activation energy found from the internal friction and that from the electrical properties, table 5.2. This is not too surprising because individual ion jumps in a glass network require somewhat different activation energies and the empirically determined values will be some average of the individual values, but the kind of average may be different for different macroscopic physical properties. There is another rather more subtle difference between the electrical and mechanical relaxation behaviour. In simple glasses, sodium ion-jumping gives rise to single absorption peaks for both electrical and mechanical losses and in both cases we can calculate a single relaxation

142

time, corresponding to a particular temperature, from the maximum in the absorption peak, i.e. from the top of the peak. However, the complete peaks are in fact much broader than would occur for a relaxation process with a single relaxation time. In order to account for the observed breadth of the peaks a spectrum of relaxation times is required. The spectra of relaxation times for sodium ion-jumping in a given glass deduced from dielectric loss and from internal friction are not identical ; it appears that the same range of ions do not contribute to both processes but a detailed interpretation has not yet been attempted.

Method	Internal friction	d.c. conductivity	a.c. losses
Activation energy in eV	0·92	0·71	0·71

Table 5.2. Comparison of activation energies for 20 mole % Na_2O–SiO_2 glass.

At low stresses, or strains, the delayed elastic strain in glass obeys the Boltzmann superposition principle (compare section 3.13). A long fibre in torsion can be used to demonstrate a mechanical ' memory ' effect analogous to the electrical one illustrated in fig. 3.22. If the fibre is twisted (several complete rotations of the end may be required, depending on the length and diameter of the fibre) first in one direction and then in the opposite one, it will, when released, show a slow change in angle of twist first one way and then the other before finally settling at zero twist.

Some recent experiments with high strength glass fibres have shown that the superposition principle breaks down at strains $\gtrsim 3$ per cent. The magnitude of the delayed strain is no longer proportional to the stress and the rate of change of strain with time, following the application of a load, is not the same as the rate of change which occurs after the removal of the load.

5.6. *Hardness of glass*

Glasses are often described as being ' hard ' or ' soft ' depending on their softening point (see fig. 1.12, page 20). In this sense a hard glass is one which has a relatively high softening point : Pyrex is a harder glass than the soda–lime–silica glasses used for windows or for bottles. In more general usage, however, the term hardness applied to a material denotes a mechanical property, e.g. the difficulty of sawing or drilling the material, its resistance to scratching or to penetration by a sharp object, and it is with hardness in this sense that we are concerned here.

143

Hardness is not a fundamental property of a material; it depends in a complex way on several basic properties combined in different ways in, for example, scratch hardness and penetration hardness. A hardness scale based on the ability of one material to scratch another was developed, originally for use with minerals, by Mohs. Minerals were ranked so that any one would scratch all those below it, but not those above it and the hardness was designated by an arbitrary series of numbers, table 5.3. In engineering the most frequently used, and

Mohs' scale			Vickers' hardness numbers	
Talc	1	Lead	Aluminium	15
Gypsum	2	Annealed aluminium	Perspex	22
		Finger nail		
Calcite	3		Mild steel	150
		Copper coin		
		Mild steel		
Fluorite	4		Tungsten	350
Apatite	5		Glasses	
		Silicate glasses	Silica	710
Feldspar	6		Aluminosilicate	590
		Tool steel (e.g. a file)	Borosilicate	580
Quartz	7		Soda–lime–silica	540
Topaz	8		Lead-silicate	450
Corundum	9		Tungsten carbide	2400
		Tungsten carbide		
Diamond	10			

Table 5.3. Hardness scales.

most generally useful, hardness tests involve indentation. A small hard ball or a pyramid is pressed into the surface of the material using a known load large enough to leave a permanent indent and an empirical hardness number is calculated from the load and the size of the indent. Two indentation tests commonly used in metallurgy are the Brinell and the Vickers hardness tests. In the Brinell method a 10 mm diameter steel or tungsten carbide ball is pressed into the surface by a measured load so that, usually, an indent 2–5 mm diameter is produced, the Brinell hardness number is taken to be the load applied divided by the surface area of the depression which remains when the ball is removed. In the Vickers method a pyramid-shaped diamond indenter, with the angle between opposite faces of the four-sided pyramid equal to $136°$, is pressed into the surface by a known load; the Vickers hardness number is taken as the weight supported by the indenter

144

(in kgf) divided by the surface area of the indent (in mm²), thus†

$$VHN = \frac{2P \sin 68°}{d^2},$$

where P is the weight supported in kgf and d in millimetres is the mean length of the diagonals of the square impression. There is a very good correlation for many metals and alloys between the Vickers hardness number and the yield stress, σ_f, i.e. the tensile stress at which permanent deformation of a macroscopic specimen first occurs,

$$VHN \simeq 2.7\sigma_f.$$

Hardness tests on glass

Silica glasses are very hard materials ; experience suggests that very few other materials will leave a scratch mark on a glass surface which is visible to the naked eye. When a hardened steel ball is pressed into the surface of a glass in a Brinell-type test a crack rather than an indentation is produced. However, a diamond pyramid will penetrate into the surface of glass and leave a small permanent indent, fig. 5.13. The Vickers hardness numbers for glasses, in table 5.3, confirm our subjective impression that these are relatively hard materials.

When a ball is pressed onto the flat surface of a block of material, the initial elastic distortion of both the ball and the surface layers of the block produces a circular area of contact. Immediately under this contact area, the material of the block is in compression but the surface surrounding the contact area is in tension, fig. 5.14 *a*. As the load on the ball is increased, the area of contact and the magnitude of the compressive and tensile stresses increase until, for metals, the stresses become large enough to produce plastic flow ; thereafter further increase in load just increases the contact area which is now formed by permanent deformation of the surface around the ball. Such permanent deformation has not been detected with glasses, instead at a critical applied load, a circular crack forms in the annular ring under tension and then further increase in load causes the crack to extend downwards and outwards into the block of glass to yield a fracture surface which has the shape of a truncated cone. The critical load decreases as the size of the ball is reduced ; for a ball of 0.5 mm diameter the critical load is of the order of 100 N. When the ball is subjected to a tangential force as well as the one normal to the surface, so that it slides across the glass, much higher tensile stresses appear in the glass surface, fig. 5.14 *b*. If the normal load is sufficiently high, a series of horseshoe-shaped cracks is produced in the glass along the track of the sliding ball, fig. 5.15. Under these conditions, the critical

† 1 kgf is approximately 9.80 N.

145

normal load required to produce cracks is very much lower : for a
3 mm diameter ball sliding across a glass surface the critical load is only
a thousandth of that required to cause cracking in the absence of the
tangential force.

By contrast when a very small ball or a hard stylus is very lightly
loaded and drawn across a glass surface it can leave a smooth furrow

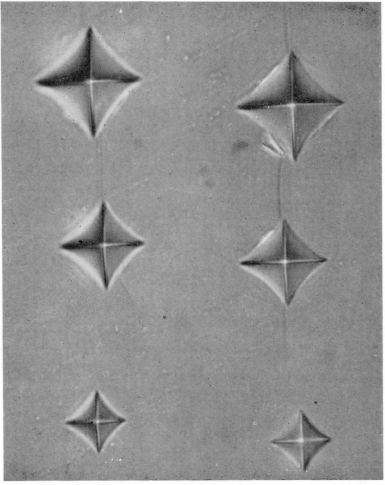

10 μm

Fig. 5.13.   Diamond pyramid indentations in a glass surface.   The three
pairs of indentations from top to bottom were produced by loads of
0·15, 0·10 and 0·05 kgf, respectively.

146

with no sign of cracking. Where two such furrows cross, the first one appears to be filled with material displaced during the formation of the second one, fig. 5.16. This suggests that the furrows are formed by lateral flow of glass around or from under the stylus in much the same way, but on a smaller scale, as the smooth scratches produced on an aluminium surface by a sharp nail or an engineer's scriber.

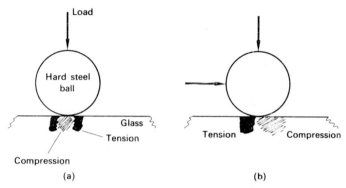

Fig. 5.14. Distribution of stresses under a spherical indentor. (*a*) With normal load ; (*b*) with normal load and a tangential force.

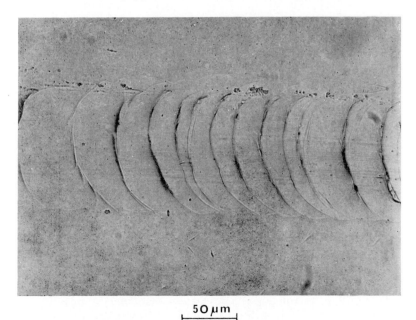

50 μm

Fig. 5.15. Horse-shoe shaped cracks in the surface of glass produced by a small ball, sliding from left to right.

147

These observations are of considerable practical interest for the understanding of the processes involved in grinding and polishing glass and also because of the dramatic effect of abrasions on the breaking strength of glass, described in Chapter 6.

Glass can be removed from a surface, a plate reduced in thickness or the shape of an article changed, by grinding with an abrasive wheel or with a slurry of carborundum power (SiC) in water. Glass is removed during grinding because the load on the small, hard, abrasive particles ' sliding ' across the surface exceeds that required to cause cracking ; as the action progresses, the surface cracks link up yielding small fragments which are removed from the surface.

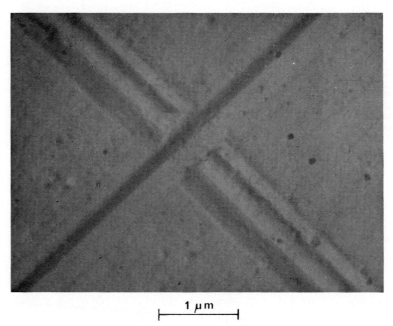

$\vdash\!\!\!-\!\!\!\overset{1\,\mu m}{\underline{}}\!\!\!-\!\!\!\dashv$

Fig. 5.16.   Crossed scratches showing the filling of the first furrow (E. Marriott).

Polishing is carried out by rubbing a slurry of rouge (ferric oxide) over the glass surface with a soft pad of felt, leather, or pitch and this process produces an optically smooth surface. The mechanisms involved in polishing are still subject to debate ; evidently the load on the finer but softer rouge particles is less than that required to cause cracking but it is also possible that local heating of the surface due to friction contributes to the flow of the surface layers. Under some polishing conditions, chemical interactions may also play some role ;

ion exchange at the surface may produce a soft layer of silica gel which is smeared out and then dried.

Although there is no sign of macroscopic plastic deformation before fracture even for very strong specimens of glass, the formation of smooth furrows and indentations in the surface does suggest that, at least over microscopic regions, plastic flow can occur in silica-based glasses at temperatures far below their softening points.   Indentations at least, indicate *plastic* flow, rather than ordinary viscous flow with an ultra-high coefficient of viscosity, because the constancy of the ratio-load/(surface area of indent) implies that there is a critical yield stress associated with the process.   However, the significance of these observations is still a matter of some controversy : there is some evidence, particularly in the case of ' pure ' silica glass, to support the view that indentation by a Vickers diamond is produced by compacting the glass under the indenter to form small regions of high-density material similar to that produced by very high hydrostatic compression.   On the other hand, as we shall see in the following chapter, there is also circumstantial evidence that plastic flow on a submicroscopic scale may be involved in the fracture of glass.   Indentation provides the most direct and also probably the least ambiguous evidence that, even at low temperatures, permanent changes in the shape over microscopic regions of a sample of glass can occur under an applied stress which exceeds a critical value.   It is possible that indentation involves both compaction and plastic flow, the proportions depending on the nature of the glass, although the mechanisms responsible for the displacement of the atoms in the network are as yet unknown.

CHAPTER 6
the fracture of glass

6.1. *Introduction*

THE fragility of glass is notorious ; if similar bowls, one of aluminium and another of glass, are dropped on to a hard floor, no-one will be in any doubt about which is more likely to be fit for further use. The trouble with glass is that below its softening point it is brittle ; it does not plastically deform to any significant extent under an applied load. The glass bowl, in contrast with the aluminium one, does not dent at the point which first strikes the floor and therefore the forces of the impact usually will be spread over a very small area, so producing high stresses. The brittle behaviour of glass is also evident from the simplest static tests : an ordinary glass rod, for example, shows no sign of any permanent deformation in tension or in bending at any stress up to the value at which fracture occurs and even then the broken pieces can be fitted together to reproduce the shape and size of the original rod. The tensile stresses required to fracture an ordinary glass and an aluminium rod are in fact much the same although the mode of fracture is very different. An aluminium rod fails in a ductile manner, i.e. it plastically deforms, forming a ' neck ' before rupture occurs, fig. 6.1.

Brittle fracture is of tremendous practical importance. On the one hand, although most metals and alloys fracture in a ductile fashion in simple tension, many of those used in structural engineering are found in service to undergo brittle fracture at stresses well below the normal tensile limit. Catastrophic failures of bridges, storage tanks, pressure vessels and pipes, and even ships and aircraft, due to the unexpected, sudden, brittle fracture of a component are still common enough to illustrate in a dramatic way the need for a better understanding of the conditions under which brittle fracture can occur in these materials. On the other hand, many technological processes depend on the ability to subdivide materials by fracture. It is much easier to split timber or rocks than to saw them ; stone-age man learnt to control the fracture of flint to produce axe-heads and knives, and many modern industrial processes depend on being able to produce very fine particles by repeated fracture.

In contrast with many crystalline materials, the fracture behaviour of glass is relatively straightforward : glasses are always extremely brittle below their softening points ; there are no complications arising from anisotropy or intercrystalline boundaries, neither does dissolved

' impurity ' produce any pronounced changes in behaviour. Indeed the general fracture characteristics of glasses are independent of their compositions and can be accounted for in terms of the behaviour of ' an elastic solid ' without invoking directly any specific atomic-scale processes. For most of this chapter there will be no need to refer to the atomic arrangement at all ; for example, we shall find that the observed strength of glasses is very much less than the theoretical ultimate strength but the explanation of this discrepancy is to be found in terms of surface flaws which are large compared with atomic irregularities and are produced by disruption of the normal surface.

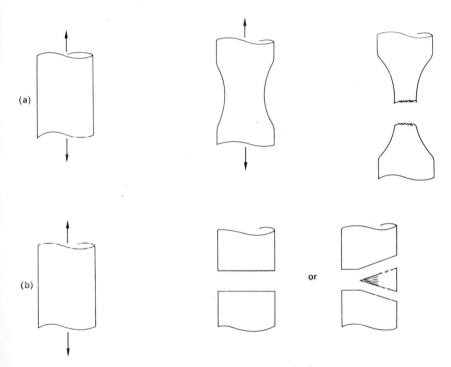

Fig. 6.1. (a) Ductile and (b) brittle fracture.

Brittle fracture occurs by the formation and propagation of a crack, but only recently have the phenomena associated with the propagation rather than the initiation of cracks, particularly in glasses, received much attention. It is likely that the fundamental problems revealed by these latest investigations are related to the details of the atomic structure but, as we shall see, at the present time only rather vague, phenomenological interpretations are available. The fundamental atomic processes involved in the propagation of a crack through a glass are still unknown.

6.2. *Theoretical strength of solids*

When an isotropic, homogeneous, brittle material, such as glass, fractures, the new surfaces formed are perpendicular to the direction of the maximum tensile stress and hence fracture involves the separation of the atoms which lie in adjacent planes perpendicular to the maximum tension, fig. 6.2 *a*. The stress required to do this is related to the attractive forces between the atoms and can be calculated if the arrangement of atoms and the forces between them are known. An applied tensile stress will increase the interatomic spacing in the direction of the stress and the curve relating the extension and the tensile stress must be of the form shown in fig. 6.2 *b*, which is of the same general shape as fig. 2.5 *a*. The fracture stress will be σ_{th}. The slope of the initial linear portion and the position of the maximum in the curves for uniaxial tension and for hydrostatic tension will be different but in principle the actual curve for uniaxial tension can be calculated from the variation in the forces between the individual atoms in the material. In practice the accuracy of the calculation is limited by the difficulty of ascertaining exactly how these forces vary at the larger separations.

A rather crude estimate of σ_{th} can be made quite simply using an argument first proposed by E. Orowan and this is adequate to reveal the major discrepancy which exists between the theoretical and observed strengths. Because the argument involves relating the maximum stress between the adjacent planes of atoms to the surface energy of the material, it also has the advantage of being applicable to all solids irrespective of the nature of the bonding and the arrangement of atoms. Orowan's argument is that the area under the stress–extension curve of fig. 6.2 *b* ($= \int \sigma \, dx$) represents the work done per unit cross-sectional area of the solid by the applied force and, when the planes of atoms have been separated completely, this work must be at least equal to the energy of the new surfaces formed.

The area under the curve is estimated by assuming that it may be represented, approximately, by the first part of a sine curve ; i.e. that

$$\sigma = \sigma_{th} \sin\left(\pi x / R\right).$$

where $x = d - d_0$ is the extension.

Then the work done in separating two adjacent planes of unit area is

$$\int_0^\infty \sigma \, dx = \int_0^R \sigma_{th} \sin\left(\frac{\pi x}{R}\right) dx = \frac{2\sigma_{th} R}{\pi}.$$

If T is the surface energy per unit area, the work done must be equal to $2T$ because unit cross-sectional area will produce two surfaces each of unit area when separation occurs.

152

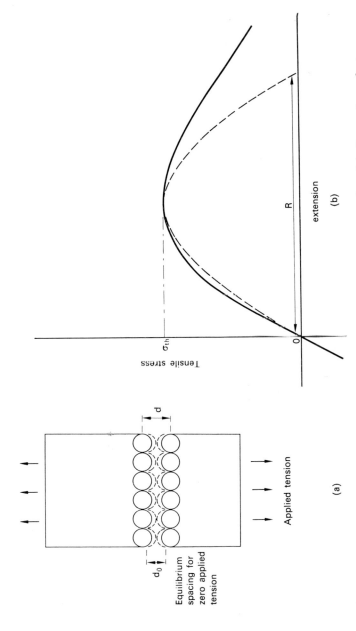

Fig. 6.2. Illustrating the calculation of the theoretical breaking strength. (a) Separation of atoms on adjacent planes; (b) variation of extension with applied tensile stress.

153

Thus

$$\frac{2\sigma_{th}R}{\pi} = 2T \quad \text{or} \quad \sigma_{th} = \frac{\pi T}{R}.$$

For small extensions, $x \ll R$,

$$\sigma = \sigma_{th} \sin \frac{\pi x}{R} \simeq \sigma_{th} \frac{\pi x}{R}$$

and Hooke's law is obeyed, so that

$$\frac{\sigma d_0}{x} = Y \quad \text{(Young's modulus)}.$$

Hence

$$R = \sigma_{th} d_0 \pi / Y$$

and

$$\sigma_{th} = (YT/d_0)^{1/2}. \tag{6.1}$$

Some values of σ_{th} for different solids calculated from equation (6.1) are shown in table 6.1, together with the tensile breaking strengths determined experimentally. Clearly there is something drastically wrong. The experimentally measured breaking strengths are two orders of magnitude, or more, lower than the theoretical estimates. These differences are much too large to be attributed to the approximations used in deriving equation (6.1) and, in any case, the much more refined calculations which have been carried out for selected solids produce very similar values for σ_{th}.

Material	Theoretical strength from equation (6.1)/MN m^{-2}	Practical strength/MN m^{-2}
Diamond	200 000	~1800
Graphite	1400	~15
Tungsten	86 000	3000 (hard-drawn wire)
Iron	40 000	2000 (high-tensile steel wire)
MgO	37 000	100
NaCl	4300	~10 (polycrystalline samples)
Silica glass	16 000	50 (ordinary samples)

Table 6.1. Theoretical and observed strengths.

The low strengths observed in practice for ordinary samples of solids are due to the presence of flaws in the structure. In some crystalline solids, particularly metals, linear faults (dislocations) in the otherwise regular atomic packing enable some planes of atoms to slide over one another under very low shearing stresses. Even under an applied tensile stress, planes at an angle to the tensile axis will experience a shearing stress and a crystal will deform plastically as microscopic blocks of atoms slide over one another rather like the individual cards in a sheared stack. This is what happens at atomic level when the stress on a metal exceeds the yield point and necking occurs in tension. By contrast, brittle materials are those in which large-scale plastic flow does not occur ; in glass there are no regular planes of atoms on which sliding could occur. Solids of this kind are severely weakened by *sharp notches* or *cracks*. As we shall describe in detail later, the stress on the material close to the tip of a crack is very much higher than the average stress over the whole cross section of a specimen so that bonds may be broken at the tip and then fracture will take place progressively across the specimen.

It is possible to produce specimens of some solids which are free from these known types of flaw, and the breaking strength of such specimens

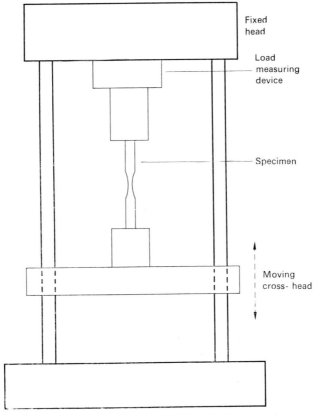

Fig. 6.3.   Schematic illustration of a testing machine.

approaches the theoretical estimates. This is particularly simple to do in the laboratory with glass but before we describe these experiments we shall review the experimental observations on ordinary samples of glass and discuss the nature and origin of the cracks and other flaws responsible for the low strength of normal samples.

6.3. *The strength of glass—experimental observations*

Methods

The stress–strain characteristics, the yield point and/or the breaking strength of materials are usually measured in a 'testing machine', fig. 6.3. The sample is held at the ends for a tensile test and the moving cross-head is driven either by a screw-thread or hydraulically so that the sample is stretched. The load is measured by a suitable device attached to the fixed head of the machine.

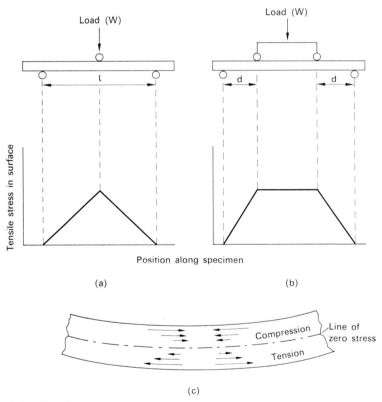

Fig. 6.4. Bending of beams. (*a*) and (*b*) show 3-point and 4-point bending, respectively; (*c*) shows the distribution of stresses within the cross section of a bent beam.

Because of the difficulties of aligning and gripping without causing fracture in the grips, very brittle materials are very often tested by bending laths or rods. A beam loaded in bending has tensile stresses on one side and compressive stress on the other, fig. 6.4. Maximum stress occurs in the surface : in 3-point bending, fig. 6.4 *a*, the maximum tensile stress occurs at a point opposite the central load and in 4-point bending, fig. 6.4 *b*, the whole of the surface between the central loading edges, on the convex side of the beam, experiences the same maximum tensile stress. Provided that the spacing of the loading points is large compared with the depth of the beam and that the deflection of the beam is small, the maximum tensile surface stress is given by

$$\sigma_{max} = \frac{M}{D},$$

where M is the maximum bending moment ($= Wl/4$ for 3-point loading and $Wd/2$ for 4-point loading) and D depends on the dimensions and

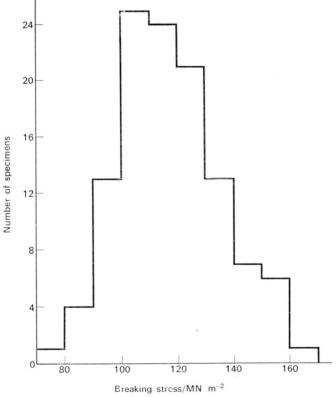

Fig. 6.5. Histogram of the breaking strength for samples of sheet glass.

157

shape of the cross section of the beam, $D = \frac{1}{4}\pi r^3$ for a circular cross section and $\frac{1}{6}$ *breadth* $\times (depth)^2$ for a rectangular section. This maximum stress is usually taken to be the breaking strength of the sample.

Experimental observations—normal samples

The simplest experiments to measure the breaking strength of glass specimens will reveal some of the main features of the behaviour. The low values of breaking strength for ordinary samples, such as rods or strips of sheet or plate glass and the large scatter in results is immediately obvious, fig. 6.5. The mean breaking strength varies with the size of specimen : if the distance between the inner loading points in 4-point bending is increased significantly, the mean strength obtained for a group of specimens decreases, while small diameter rods or fibres show higher strengths than large rods, table 6.2. Under prolonged loading, samples will fail eventually at a fraction of the stress required to produce immediate failure, fig. 6.6. This effect, known as *static*

Length/mm (diameter $\sim 13\ \mu$m)	Tensile breaking strength/MN m^{-2}
5	1470
10	1200
20	1200
90	750
1560	700

Table 6.2. Effect of length of sample on breaking strength. (These fibres were damaged in the course of normal handling.) (After F. O. Anderegg).

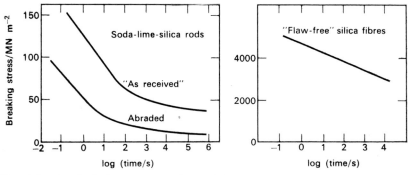

Fig. 6.6. Static fatigue or the effect of duration of loading on the observed breaking strength.

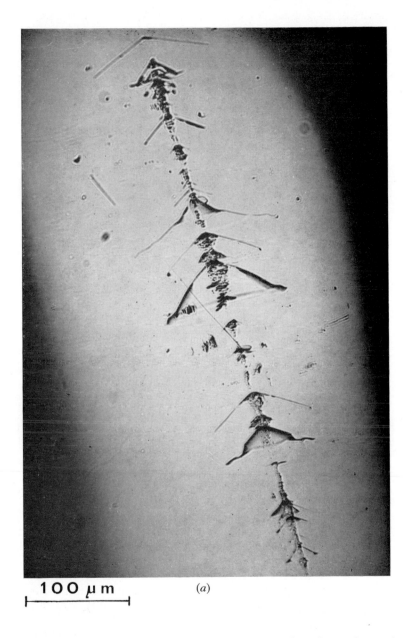

100 μm   (*a*)

Fig. 6.7. (*a*) Surface damage produced by light abrasion of one glass rod on another.

159

100 μm   (b)

Fig. 6.7. (b) End of a broken rod showing the area, on the original surface, where fracture initiated from an abrasion crack. (The visibility of the fine surface cracks has been enhanced by a very short etch in dilute hydrogen fluoride).

fatigue, also results in a change in the average breaking strength with rate of loading.

The breaking strength of glass is particularly sensitive to abrasion of the surface : the strength of freshly drawn fibres drops sharply if they are handled or are allowed to rub against one another. Glaziers and glass-workers make use of this sensitivity to induce fracture at a chosen position by scratching the surface with a diamond or other very hard material.

Examination of characteristic features which occur on fracture surfaces can be used to locate the fracture origin (see section 6.7). These indicate that fracture invariably starts from the surface even for samples loaded in tension so that the whole cross section is under uniform stress. For ordinary samples, the origin of fracture is usually in one of the areas which have been damaged by abrasion, fig. 6.7. As we saw in the previous chapter (section 5.6), abrasion can produce small cracks in the surface of glass and we can anticipate that a cracked specimen will break at a much lower stress than an intact one. The low strength, high scatter and the variation with specimen size normally observed are explicable, in principle, in terms of such surface flaws. For a sample under uniform stress, fracture will occur at the ' worst ' flaw but the severity of the ' worst ' flaw may vary from one sample to another and the smaller the specimen, or the smaller the surface area under high stress, the less the chance of its including a very severe flaw and the stronger it is likely to be.

Untouched samples

The strength of samples of glass free from abrasion flaws can be determined by heating and necking down the central section of a rod and then breaking the narrow central fibre either in tension, or in bending as in fig. 6.8, so that no contact is made with the freshly drawn, fire-finished surface. Such specimens are significantly stronger although, usually, they are still far below the theoretical strength. The strengths still show high scatter and vary with the size of the specimen and rate of loading, and again failure starts from the surface.

For many years it was believed that the size-effect and scatter in the strength of fibres which had not been mechanically damaged in any way, was of fundamental significance. The behaviour was attributed, despite the total lack of direct evidence, to the presence of very fine cracks which were in some way inherent in the structure of glasses.

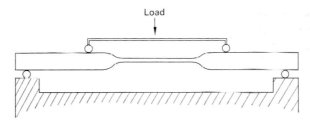

Fig. 6.8. Loading arrangement suitable for testing ' flaw-free ' samples : the central, drawn section remains untouched as the sample is loaded.

The original idea, that a random distribution of flaws which amplified the applied stress might account for the variable low strength of many solids, was due to A. A. Griffith in 1920. He also showed that small cracks were one kind of flaw with this property and suggested a criterion for the extension of such a crack. The general significance of this criterion, which we shall describe later, seems to have been submerged for many years. Fundamental work on very brittle materials tended to be concerned with attempting to reveal the origin or the distribution of the ' inherent cracks ' which were supposed to exist in these materials, while, for the more ductile solids, especially the metals, the processes giving rise to plastic deformation received most of the attention.

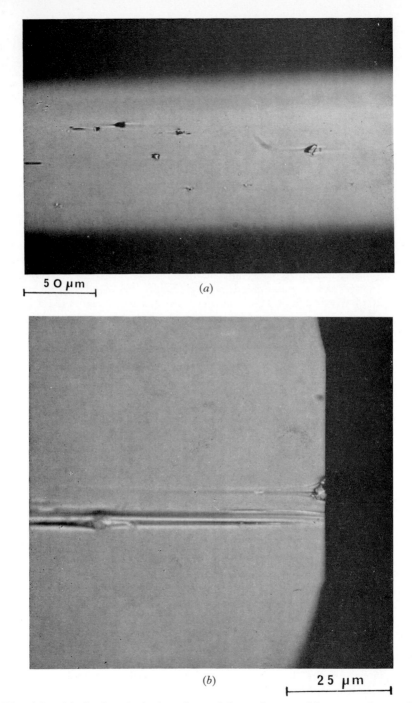

5 O µm   (a)

(b)   2 5 µm

Fig. 6.9. (a) Surface inclusions formed from dust particles on a glass rod ;
(b) end of a broken rod showing the area on the original surface where
fracture initiation occurred from a surface inclusion.

The flaws responsible for the low strength of fresh, 'untouched' surfaces of fibres have been identified as small, usually microscopic, particles of foreign matter which became bonded to the surface while it was hot, fig. 6.9. The possibility that flaws of this kind would weaken a solid was recognized by Griffith and it has long been known that larger foreign particles, e.g. undissolved batch or lumps of refractory from the furnace, could cause fracture at low stresses. However, the microscopic surface inclusions have only been quite recently identified and the surface shown to be very strong in between these flaws. Glass is an excellent solvent at high temperatures ; silica glasses can be formed with a wide variety of inorganic compounds and this is perhaps related to the ease with which small inorganic particles, e.g. in air-borne dust, will bond firmly to the surface of all types of silica glass even at temperatures as low as 200°C, well below the softening point.

Abrasion cracks and surface inclusions provide the two serious strength-impairing flaws on the surface of commercial glasses ; typically, the abrasion damage resulting from normal handling results in failure at about 50 MN m^{-2} and heat treatment at the annealing temperature in the laboratory (a considerably less dusty atmosphere than a glass works !) gives surface inclusions which cause failure at about 400 MN m^{-2}. These strength levels are much the same for glasses of different composition. Both abrasion damage and surface inclusions have been shown to initiate fracture in a number of other very brittle solids. Figure 6.10 illustrates the effect of abrasion damage on the breaking strength of sapphire crystals (Al$_2$O$_3$).

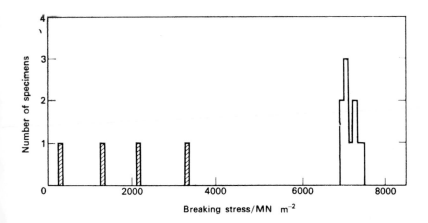

Fig. 6.10. Effect of surface abrasion on the strength of sapphire single crystals ; the surfaces of all the crystals were fire-polished and then some were rolled over carborundum powder before breaking (shaded results) (after B. A. Proctor).

'Flaw-free' samples

Glass specimens free from both these kinds of surface flaw can be produced quite simply in the laboratory: abrasion cracks can be avoided by not touching the freshly drawn surface, and surface inclusions can be avoided, or at least drastically reduced in frequency, by drawing fibres very quickly from a pool of glass which is at a high temperature. If the temperature is sufficiently high any contaminating particles will dissolve completely. Rapid drawing will minimize the chance of air-borne contamination reaching the surface while it is still hot. Alternatively, both types of flaw can be removed by slowly dissolving away the surface of the glass with dilute hydrogen fluoride (HF). The attack must be carefully controlled, so that a highly polished surface is produced if reproducible results are to be obtained and this is much easier to achieve with some glass compositions than with others. Carefully prepared (and tested !) specimens of either of these kinds show very reproducible strengths; at room temperature and normal rates of loading, soda–lime–silica glass fails at 3400 MN m^{-2} and silica at 6000 MN m^{-2}, fig. 6.11. The scatter in the results is within the experimental error associated with the measurement of load and specimen size, but these 'flaw-free' samples still exhibit some static fatigue (fig. 6.6 b) and the strengths are still somewhat below the best theoretical estimates based on an ideally brittle model. These are, however, remarkably high strengths and before describing the mechanisms by which the surface flaws reduce the strength of normal samples, it may be appropriate to discuss briefly the possibility of utilizing the high strength of 'flaw-free' glass.

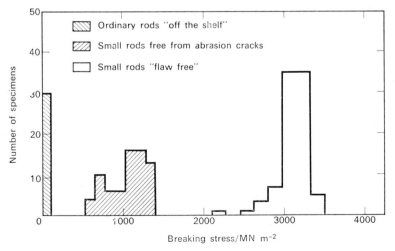

Fig. 6.11. Comparison of the breaking strengths of 'ordinary', 'untouched' and 'flaw-free' glass rods.

164

Material	Practical strength/MN m^{-2}	Relative density	$\dfrac{\text{Strength}}{\text{Rel. density}}$	$\dfrac{\text{Young's mod.}}{\text{Rel. density}} : 10^{-4}/\text{MN m}^{-2}$
Al alloy	300	2·8	107	2·5
Mild steel	460	7·8	59	2·6
Steel wire (high tensile)	2000	7·8	256	2·6
Plastics	~100	1→2	50→100	0·15→0·3
High-tenacity nylon or rayon (tyre cord)	~1000	~1·3	760	~0·3
'Flaw-free' silica glass	5000	2·2	2270	3·2
'Flaw-free' 'soft' glass	3500	2·2	1600	3·2
Carbon fibre (graphitized textile fibres)	3200	1·9	1700	26

Table 6.3. Strength and stiffness per unit mass for various materials.

Applications of high strength glass

In many contemporary applications, from rocketry to boat-building, the strength per unit mass of the structural materials is of primary interest and on this basis ' flaw-free ' glass stands out as a particularly attractive material, table 6.3. Although there should be no insuperable difficulties in preparing large pieces of glass which have very high breaking strengths, say of order 2000–3000 MN m^{-2}, in practice there is no real prospect of using glass in bulk form for load-bearing or structural purposes ; the susceptibility to surface damage and the low resistance to crack propagation make it inherently unsafe.

A bundle of fibres is much safer than a single bar of glass, for if one fibre is damaged and breaks under the tensile load, the crack does not propagate into the neighbouring fibres. The high strength of glass can be utilized by embedding fibres in a suitable matrix. The matrix serves to separate the fibres, preventing both mutual and external abrasion of their surfaces ; it also transfers the applied stress to the fibres. For this purpose the matrix material needs to be either relatively soft and ductile, like aluminium, or to have a low elastic constant and a high fracture strain, like many polymers ; in either case the applied stress is transferred to the embedded lengths of fibre by the shear stresses at the interface between fibres and matrix. This is the mechanism which operates in ordinary glass-fibre reinforced plastics. Normally, however, very strong ' flaw-free ' fibres are not used in making these materials. Processes have been developed in which a film of aluminium or of thermosetting resin is applied to the fibre as it is drawn, so that the surface is protected almost as soon as it is formed. Bundles of these coated fibres can then be moulded to form a composite which has a very high volume fraction of glass fibre. If the fibres are all aligned in one direction, a very high tensile breaking strength approaching that of ' flaw-free ' glass can be achieved in the direction of the alignment. Although the strength in the transverse directions will be very much lower, for some applications this is not a serious problem. There is, however, still one major snag with high-strength glass composites : they are not stiff enough for many applications.

The tensile elastic modulus of a fibre-reinforced composite material in the direction of fibre alignment depends on the volume fraction of fibres, but it will not exceed that of the fibres, even for very high volume fractions. If a breaking strength approaching that of the fibres is achieved in such a composite then the elastic constant, and hence the elastic strain at which failure of the material occurs, will be comparable to that observed for individual fibres. Thus for a glass-fibre reinforced material which realizes the full high strength of the glass, the elastic strain at failure will be about 5 per cent. Such large failure strains have some advantage for fishing-rods and vaulting-poles but are intolerable in the structural members of a bridge, a large microwave

166

dish aerial or an airplane wing. It is largely because of the much higher Young's modulus of the fibres that recent interest in high-strength composites has turned towards carbon-fibre reinforced materials, compare table 6.3.

6.4. *Stress-raisers and cracks*

We can deduce from the experimental observations on the breaking strength of glass samples prepared and tested in various ways that it requires a very high stress to initiate a crack but that, once formed, it will propagate at quite low stresses. In this section we shall describe the mechanisms responsible for initiating cracks and consider the condition for their propagation.

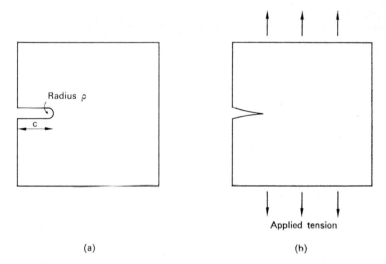

(a)   (b)

Fig. 6.12. (*a*) A notch and (*b*) a crack in the edge of a thin sheet.

Abraded samples of glass already have small cracks in the surface, formed under the high but very local stresses developed by quite small loads applied to minute areas of the surface. The breaking strength of abraded specimens is determined therefore by the condition for the propagation of these preformed cracks. With 'untouched' samples, fracture starts under low applied stress at a surface inclusion. There-fore, either an inclusion makes crack initiation under an applied stress much easier or inclusions already have cracks associated with them. In fact, for different types of glass and contamination both mechanisms have been observed. A foreign particle which is stuck to the surface of glass when it is hot, or even dissolved so as to form a small glassy patch of very different composition from the bulk of the glass, may

167

cause large stresses to develop during cooling because of the difference in the thermal contraction between the particle or patch and the surrounding glass. Under some circumstances these stresses alone are large enough to initiate cracks in the glass. It is also possible for the inclusion to act as a *stress-raiser*, i.e. to amplify the applied stress, and this is probably the more common mechanism by which naturally occurring inclusions weaken glass.

Notches

The simplest type of ' inclusion ' is a hole or a notch, in which case the intruding ' foreign particle ' is air. The effect of such large-scale flaws as rivet-holes, port-holes, screw-threads, etc., on the stress distribution is of considerable practical importance in engineering. Not surprisingly, therefore, many theoretical and experimental investigations have been carried out on this problem by engineers.

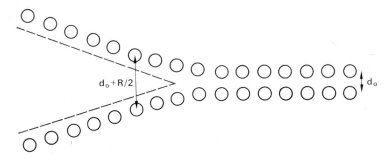

Fig. 6.13.   Arrangement of the atoms near the tip of a crack.

One of the first detailed calculations of the stress-raising effect of a notch was due to C. E. Inglis in 1913. He showed that the stress near the tip of a notch could be very much larger than the average stress, i.e. (applied load/cross-sectional area). A notch, depth c, with a semicircular tip, radius ρ, in the edge of a thin elastic sheet to which a tensile stress σ is applied, fig. 6.12 a, produces a stress σ_m close to its tip given by

$$\sigma_m = \sigma[1 + 2(c/\rho)^{1/2}] \simeq 2\sigma(c/\rho)^{1/2} \quad \text{for} \quad \rho \ll c. \quad (6.2)$$

For $c \sim 10^{-6}$ mm and $\rho \sim 10^{-6}$ mm, $\sigma_m \approx 200\sigma$, so that, if they are very sharp, notches do not have to be very long in order to produce, locally, large magnifications of the average stress on the sheet. The local stress near the tip of a sharp notch may reach that required to break interatomic bonds, i.e. σ_{th}, even though the average applied stress is quite low.

Cracks

In order to deal with the effect of atomically sharp cracks, rather than notches, a different approach is required. Equation (6.2) was derived from the theory of the behaviour of an elastic continuum, which ignores the real atomic nature of matter, so that it is strictly valid only for $\rho \gg$ the interatomic spacing.

When a crack is formed under an applied tensile stress in an ideally brittle material, there are no atoms missing from within the gap but the surfaces of the crack will be held apart by the applied tension. The crack surfaces will approach one another tangentially near the tip of the crack, fig. 6.12 b, and in this region the atoms in the opposite surfaces will be close enough for attractive forces to exist across the gap. Somewhere in this region near the tip, the spacing of the atoms will correspond to the separation at which the net attractive force is a maximum, i.e. the stress at some point near the end of an elastic crack is automatically equal to σ_{th}, fig. 6.13. An applied tension must exist in order to hold the crack open against the attractive forces near the tip ; if the tension is reduced the crack will, as it were, start to zip itself up ; if the tension is increased the crack will extend.

The critical value of the tensile stress required to hold an elastic crack at a given length was first calculated by Griffith. Although it is now doubtful whether the expression derived by Griffith is applicable in its original form to any real solid, the principle which he used provided the essential theoretical framework for the analysis of practical problems of brittle fracture in engineering materials : the full potential has only very recently been realized.

To calculate the critical stress, Griffith considered the energy stored in the cracked solid. If a thin elastic sheet is stretched by an applied stress, σ, the elastic strain energy per unit volume stored in the sheet is $\frac{1}{2}(stress \times strain) = \frac{1}{2}(\sigma^2/Y)$. Then if a crack of length c is formed in the edge of the sheet, the tensile strain will be removed from a region immediately above and below the crack because the tensile forces cannot be transmitted across it. Suppose, first of all, that the sheet has been stretched to a new fixed length (constant strain) so that when the crack is formed, the ends of the sheet do not move and therefore the applied forces required to maintain the extension at the constant value do no work. In this case the stored elastic energy in the sheet is reduced when the crack is formed ; the reduction is of the order of that contained originally in the semicircular region around the crack, fig. 6.14, i.e.

$$\frac{\sigma^2}{2Y}\left(\frac{\pi c^2}{2}\right) \text{ per unit thickness of sheet.}$$

The exact calculations of the redistribution of strain energy give just twice this simple estimate and also show that the energy released

169

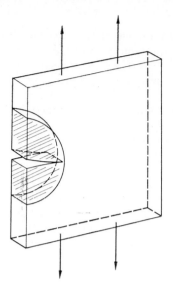

Fig. 6.14.   Illustrating the volume relieved of strain energy by the formation
of an edge crack.

is exactly the same for an edge crack of depth c, whether the sheet is
held at constant strain or at constant stress :

$$\text{energy released} = \sigma^2\pi c^2/2Y \text{ per unit thickness.} \qquad (6.3)$$

At constant stress the formation of a crack increases the stored elastic
energy but causes a displacement of the ends of the sheet so that work
is done by the applied forces : the work done is greater, by just the
expression (6.3), than the increase in stored energy in the sheet.

The formation of a crack will create additional surfaces and the surface
energy associated with these will be $2Tc$ per unit thickness for the edge
crack, where T is the surface tension.   The total increase in the energy
of the sheet due to the formation of the crack is therefore

$$-\frac{\pi\sigma^2 c^2}{2Y} + 2Tc.$$

Griffith argued that the condition for the growth of a pre-existing crack
under an applied stress is simply that the total energy of the sheet should
decrease as the crack length increases, i.e.

$$\mathrm{d}/\mathrm{d}c(2Tc - \pi\sigma^2 c^2/2Y) < 0,$$

thus

$$\sigma > (2YT/\pi c)^{1/2}.$$

170

Hence, for a sheet containing an edge crack of length c, the critical stress is given by

$$\sigma_g = (2 Y T/\pi c)^{1/2}. \qquad (6.4)$$

Expressions of the same form but with slightly different numerical factors have since been derived, using the same criterion, for the critical stress for thick sheets and various other crack geometries. Hence, from equations (6.1) and (6.4) we have

$$\sigma_{th}/\sigma_g \sim (c/d_0)^{1/2},$$

so that a material need contain only very short elastic cracks, $\sim 10^{-3}$ mm long, for its strength to fall by a factor of 100.

For an ideally brittle material, cracks formed for example by high local contact stresses should close up when the stress is removed; glass is not ideally brittle and this may explain why abrasion cracks remain in the surface as serious flaws. Other explanations are possible : that small particles of debris wedge open the end of the crack and/or that the adsorption of gas molecules onto the freshly formed surfaces decreases the interatomic forces and prevents the ' zipping up '.

6.5. Fracture energy

The important idea introduced by Griffith, which has proved so useful in the analysis of the phenomenon of brittle fracture, is that the extension of a crack occurs when the total elastic energy of the system can decrease by the process. This principle can be stated in a slightly different way and this is the form in which it is now commonly expressed : a crack will extend if the rate of release of elastic energy due to an increase in crack length (i.e. $\partial u/\partial c$) is greater than the energy required to create the new surfaces. Griffith assumed that the energy required to create unit area of new surface was just the reversible thermodynamic surface energy (equivalent to the surface tension). For real materials this is rarely the case : even for very brittle materials small amounts of plastic deformation or other irreversible effects occur at crack tips and the energy required to create unit area of new surface by propagating a crack (called the ' fracture energy ' or ' work of fracture ' and denoted by γ) is greater than the thermodynamic surface energy. However, the criterion for the extension of a crack can be retained, and the critical tensile stress is still given by an expression like the one derived by Griffith, equation (6.4), provided T is replaced by γ.

The fracture energy of a material can be measured by experiment. A particularly convenient form of specimen, known as the double cantilever beam, is illustrated in fig. 6.15 ; the grooves along the length ensure that the crack propagates down the middle of the specimen.

171

This geometry has the advantage that catastrophic propagation of the crack may be avoided by controlling the displacement of the ends of the cantilever arms and, by progressively increasing this displacement and observing the corresponding crack lengths, several independent estimates of the fracture energy can be obtained from a single specimen.

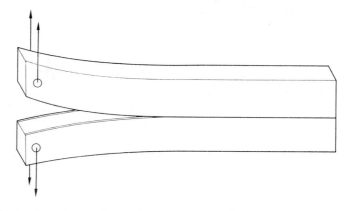

Fig. 6.15. Double cantilever beam specimen for measurement of fracture energy. Shallow grooves down the front and back faces of the specimen stabilize the direction of crack propagation.

The surface energy and experimentally measured fracture energy for several brittle materials are shown in table 6.4. The higher values of the fracture energy may be accounted for, in a general way, by supposing that there is some plastic flow in the highly stressed material at the crack tip. This can be shown to occur in many brittle metals and some polymers. Plastic, or viscous, deformation of the material just ahead

Material	Surface energy/J m^{-2}	Fracture energy/J m^{-2}
Al$_2$O$_3$	1·1	12–30
LiF	0·17	0·34
NaCl	0·15	0·31
Perspex	∼0·5	200
Plate glass (soda–lime–silica)	∼0·3–0·4	4
Pyrex	—	5
Mild steel	—	100 (at 78 K)
Cast iron	—	1000

Table 6.4. Estimated surface energies and experimental fracture energies.

of the crack tip will blunt it and in some cases the tip can be resolved with an optical microscope, fig. 6.16. The amount of deformation and extent of the deformed zone must be limited because, in practice, if the load is increased the crack extends : once a certain amount of deformation has occurred, the material prefers to rupture rather than to increase the deformation or to expand the size of the deformed zone. (Indeed it is this characteristic which makes the material brittle rather than ductile, so that an explanation of fracture behaviour in terms of

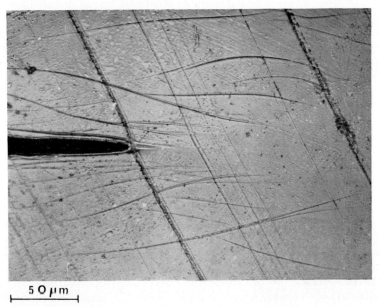

5 O μm

Fig. 6.16. A blunted crack tip in a sheet of Perspex. The deformation of a fortuitously pre-existing scratch reveals the plastic flow in the material ahead of the crack tip.

the atomic structure must explain why and predict when, this change occurs. This is not yet possible ; research in this area is still at the earlier, empirical stage for most materials.)

As the crack extends, the material in its path is deformed : the zone of deformation is pushed ahead of the crack. We can estimate the energy expended in this process if we assume that the irreversible deformation starts at a critical local stress, σ_f, and continues at that same stress until the material in the plastically deformed zone is stretched by an amount α, whereat local rupture occurs. The work done in forming the zone is then

$$force \times displacement = \sigma_f \, (area) \; \alpha$$

or the work per unit area is just $\sigma_f\alpha$. The total energy required to extend a crack so that it creates two surfaces each of unit area is then given by :

$$2\gamma = \text{surface energy} + \text{plastic work} = 2T + \sigma_f\alpha.$$

Since for most brittle materials $\gamma \gg T$ we can often write

$$\gamma = \tfrac{1}{2}\sigma_f\alpha. \qquad (6.5)$$

It is interesting to see how this works out for glass : we can use a value of flow stress calculated from diamond pyramid hardness indentations as an estimate of σ_f and then, using the experimentally measured values for γ, we can calculate a value for α which will provide an estimate of the diameter of the crack tip. It is also possible, from a more detailed analysis, to calculate the radius of the plastic zone from γ, σ_f and Young's modulus. It is found that α is about 0·8 nm and the radius of the zone is about 3 nm for a commercial soda–lime–silica glass (plate glass). Thus a very small zone of plastic flow is sufficient to account for the observed fracture energy. There is, as yet, no direct evidence that the observed fracture energy of glass is due to the occurrence of plastic flow at crack tips ; it is not obvious how a plastic zone, or a crack tip radius, of this order of size might be detected directly. However, if some kind of flow process is linked with fracture, then the estimates of the theoretical strength of glasses based on an ideally brittle model may be too high. D. M. Marsh has pointed out that the high experimental breaking strengths reported for ' flaw-free ' specimens of different glasses, tested at various temperatures, can be correlated with the corresponding flow stresses calculated from diamond pyramid indentations. He suggests that ' flawless ' glass fractures when plastic deformation occurs ; when the first zone of plastic deformation, initiated at a favourable irregularity in the atomic arrangement, reaches a critical size it ruptures and a crack is formed. The fracture of glass would nevertheless appear perfectly brittle because the flow is confined to a minute volume of material.

The strength of ' flaw-free ' glass specimens depends on the environment to which their surfaces are exposed during the breaking test, especially on the concentration of water or water vapour. In addition, as we have already seen (cf. fig. 6.6 b) specimens of this type still exhibit static fatigue. The Vickers hardness of similar glasses also depends on the environment and on the duration of load, i.e. the diamond sinks into the glass producing a progressively larger indentation if the load is left on. (The rate of change of hardness number is independent of the load and becomes very small after $\sim 10^5$ s in air, so that this ' indentation creep ' is clearly something quite different from a simple viscous relaxation). Although it is not yet understood why or how indentation creep occurs, Marsh's hypothesis would imply that there is an intimate connection between the changes in strength and in hardness number. At the present time, however, the correlation between the behaviour of these two properties must be considered only qualitative, and the evidence for a flow process being responsible for the ultimate strength of glass must be considered circumstantial, because of difficulties in calculating reliably and accurately the magnitude of the flow stress from the Vickers hardness number. The simple relationship between yield (flow) stress and VHN established for the softer metals (section 5.6) is certainly not valid for glasses ; $VHN/2\cdot7$ is significantly less than the fracture strength of 'flaw-free' specimens and no macroscopic flow has been observed before fracture. Other, more general, relationships between flow stress and hardness number have been derived theoretically but their applicability to glass is still open to question, particularly since the way in which glass deforms under a pyramid indenter is still not clear.

Detailed interpretations of the phenomena associated with fracture initiation in 'flaw-free' specimens, with the energy required for crack propagation and with plastic flow in glasses, will almost certainly require much more information on the atomic arrangement in these materials, particularly over distances of the order of 10 nm, than is presently available. Once again, therefore, it is the current lack of sufficiently detailed, unambiguous information which limits our discussion of the fundamental basis of an important property of glass.

Engineering application of fracture energy

Measurement of the fracture energy provides a sophisticated way of assessing empirically the brittleness of a material and this, or the closely related parameter called *fracture toughness*, $=(2\gamma Y)^{1/2}$, is being used increasingly in engineering. For most applications in structural engineering, materials must be tough as well as strong. It is virtually impossible to prepare materials or to fabricate complex structures on an industrial scale without introducing small cracks, from weld defects, macroscopic inclusions, etc., so that it is essential that a structural material should have both a high yield stress, to resist failure by macroscopic plastic deformation, and a high fracture energy, to resist brittle fracture by the growth of the already existing cracks.

Cracks in the components, such as large steel plates or girders, of an engineering structure may be very much larger than the microscopic ones with which we have been concerned in glass. Since the brittle-fracture strength of a particular component is determined by the size of the cracks in it, evaluating the safe loads for the individual components, and hence the complete structure, from measurements of the strength of other perhaps much smaller laboratory-scale specimens of the same materials is a very unreliable procedure. Traditionally, design engineers have dealt with this problem by using large safety factors : the complete structure is designed so that the tensile stresses in any critical component are always a small fraction of the mean strength observed in appropriate routine tests on samples of the same material. An alternative approach to the problem, now finding increasing use, is based on Griffith's criterion for the extension of a pre-existing crack. This is feasible because, for many structural materials, it is possible to ascertain, by non-destructive tests, whether there are any cracks exceeding a specified length in a particular component. If an upper limit can be set to the length of *all* the cracks in a particular component, and the fracture toughness of the material is known, the stress which that actual component will support can be calculated from the modified form of equation (6.4), i.e.

$$\sigma = \left(\frac{2\gamma Y}{\pi c_{max}} \right)^{1/2}.$$

Unlike the brittle-fracture strength, the fracture toughness is a reproducible characteristic of the material. Although, for any given material, the fracture toughness may vary with environment, rate of loading, etc., it can be measured, for whatever conditions seem relevant to a particular application, by laboratory tests on relatively small samples of the material.

Even from the standpoint of fracture toughness, the behaviour of the common and important structural materials is very complex and it would be ludicrously naive to expect that this approach will enable engineers quickly to eliminate accidents due to brittle fracture. However, it does hold much promise, perhaps not least because the fracture toughness is both a useful practical parameter and one which can in principle be interpreted in a fundamental way in terms of the atomic-scale behaviour of a material.

6.6. *Speed of crack propagation*

Once fracture is initiated in a brittle material under a uniform stress, it is completed very rapidly but it does take a finite time. Separation of the atoms in the adjacent layers does not occur simultaneously across the whole fracture surface, but progressively by the propagation of a crack and there are some physical limitations on the rate at which this can occur. That is, the speed of a propagating crack is limited. A very general upper limit must be set by the rate at which ' information ' about the position of the crack front, and therefore the adjustment of stresses and strains, can be transmitted into the unbroken part of the material. The crack certainly cannot travel faster than the mechanical waves which carry this information ; thus it cannot travel faster than the speed of sound in the material. There is a kinetic energy associated with a propagating crack, because extension of the crack requires the atoms in and near the walls to be displaced laterally ; some of this kinetic energy

176

will appear eventually as the observable velocity of the broken pieces. Normally the supply of energy limits the maximum speed of crack propagation to a fraction of the speed of sound. The following qualitative argument shows how this comes about.

When the elastic energy release rate ($\partial u/\partial c$) is greater than twice the fracture energy (2γ), growth of a pre-existing crack will occur. For an edge-crack in a thin sheet this condition gives from equation (6.4)

$$\frac{\pi\sigma^2 c}{Y} > 2\gamma.$$

As the crack grows, c increases so that if the applied stress σ is constant, the energy release rate increases and more energy is released during each increment of growth than is required for the formation of the new surfaces by fracture. The excess energy is transformed into kinetic energy as the crack accelerates. The kinetic energy associated with the crack increases with both the speed and the length of the crack : the faster the advance of the crack tip, the faster the material in the walls must be displaced and the longer the crack, the greater the length and therefore mass of material which must be displaced. Long cracks have a greater kinetic energy than short cracks *with the same speed*. Thus, although the energy release rate increases with crack length, eventually the crack may reach a speed where the energy in excess of that needed to form the new surfaces is only just sufficient to provide the increase in kinetic energy due to increasing length and there will be none left over to provide for an increase in speed.

This maximum speed has been calculated in several different ways but all the calculations involve rather severe approximations, for example that the material is ideally elastic and that the stress distribution around the tip of a crack extending at high speed is the same as that around a stationary one. The more abstract methods show that a crack in an ideally brittle material can be represented by a combination of surface waves and then it follows that the extension of the crack cannot occur faster than these waves can travel. To some extent it is reassuring that all these calculations produce rather similar answers for the maximum speed, viz.

$$[0\cdot4 \text{ to } 0\cdot6] \times (Y/\rho)^{1/2}.$$

For glass this is ~ 2 to 3×10^4 m s^{-1}. Experimental measurements on glasses and other brittle materials confirm that cracks do reach a terminal speed which is of this order, although for glasses, irrespective of composition, it is consistently less than the predicted values. It is not yet known why this should be so.

6.7. *Fracture surface morphology*

In service, the fracture of glass seldom occurs under simple, uniform tension. When fracture is induced by thermal shock, there may be

177

gradients of stress and changes in the orientation of the stress along the length, across the width and also through the thickness of the article; when a large pane of plate glass is broken by the impact of a cricket ball, cracks initiated close to the position of impact run across the whole pane under the stress produced as the momentum of the ball causes the pane to bend. Examination of fracture surfaces, even with the naked eye, reveals a variety of markings and experience has shown that detailed study of these can provide useful information on the course and progress

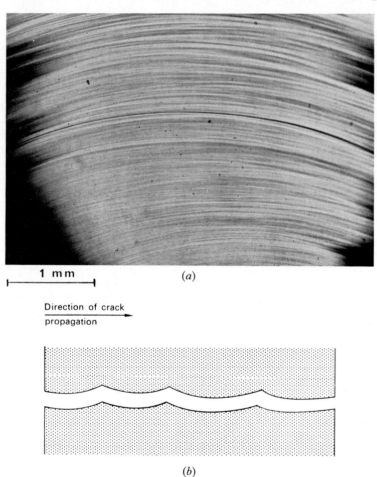

1 mm   (a)

Direction of crack propagation

(b)

Fig. 6.17. Ripples on glass fracture surfaces. (a) Rib-marks. (Direction of crack propagation is from bottom to top of the picture); (b) section showing form of rib-marks; (c) Wallner lines. (Crack origin is in lower centre of picture. Note the few prominent rib marks in upper left-hand corner.)

178

(c)

Fig. 6.17. (cont.)

of the fracture. The study of the morphology of fracture surfaces is
therefore of practical value as well as of fundamental interest.

When a crack propagates in glass under a relatively low stress, i.e.
when the elastic energy release rate is small, the fracture surfaces produced are optically smooth but they may be marked by occasional
ripples and striations. Figures 6.17 and 6.18. When the elastic
energy release rate increases, much rougher fracture surfaces are produced and the fracture may branch, fig. 6.19. This roughening and
branching lead to a characteristic pattern on tensile fracture surfaces
which is illustrated in fig. 6.24 and is discussed in detail later.

Rib-marks and striations

Smooth rippled surfaces are often produced when a flake of glass is
chipped from the edge of an article by a glancing blow. Ripples are
always curved and the convex side faces the direction in which the
crack propagates. Some of the ripples, known as rib-marks (fig.
6.17 a and b) indicate the actual profile of the crack front ; they correspond to an inflexion in the plane of crack propagation which affected
the whole of the crack front at the same time. Other ripples, known as
Wallner lines, after the man who first explained their origin, are formed
progressively as the crack propagates and these can be used to calculate

179

the speed with which crack tip passed through that region.   *Striations* are steps in the fracture surface parallel to the local direction of crack propagation and they occur when adjacent sections of a crack front follow paths at slightly different levels, fig. 6.18 *b*.   The actual step is

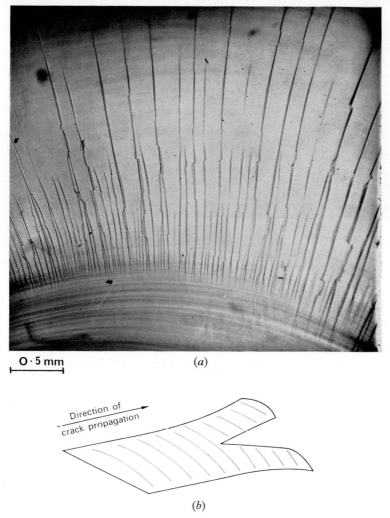

O·5 mm

(*a*)

Direction of
crack propagation

(*b*)

Fig. 6.18.   Striations.   (*a*) Photomicrograph of striations on a glass fracture
 surface.   (Direction of crack propagation is from bottom to top of the
 picture.)   (*b*) and (*c*) show the process of crack front division and overlap
 which leads to the formation of a striation.   (The direction of propagation of the original crack is indicated by the arrow in (*b*) and is normal
 to the plane of the figure in (*c*).)

180

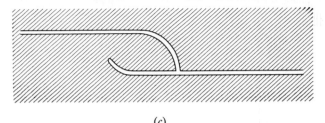

(c)

Fig. 6.18.[(cont.)

formed as the separated cracks overlap and one curves up (or down) to join the other, fig. 6.18 c.

Since there are no regular planes of atoms in glass, there are no preferred planes of fracture within the material itself and a crack propagates along a surface perpendicular to the instantaneous local maximum stress at its tip. The initial plane of propagation is therefore perpendicular to the direction of the maximum applied tension. However, the growth of a crack may change not only the magnitude but also the orientation of the stresses in the unbroken parts of the body, so that the prediction of the complete path of a crack is a complex dynamic problem. In non-uniform stress fields, cracks often follow curved paths and this produces the commonly observed *conchoidal* (i.e. shell-like) fracture surfaces. If the direction of maximum tensile stress in the material just ahead of a crack tip is rotated in a plane parallel to the direction of propagation, ABCD in fig. 6.20 *a*, then the crack can curve continuously out of the original plane. If the axis of the maximum tensile stress is rotated in a plane perpendicular to the direction of propagation, PQRS in fig. 6.20 *b*, continuous adjustment of the plane of propagation, requiring a twisting of the entire crack front, is not possible. In this case, the original crack front breaks up into separate small cracks which can adjust to the new orientation and these separated cracks produce striations.

The origin of rib-marks and Wallner lines

Ripples in a smooth fracture surface are produced by transitory changes in the direction of maximum tension in the material at the crack tip, brought about by elastic stress pulses. Stress pulses or stress waves in a solid are analogous to the pulses and waves which can be produced in an open-coiled spring (e.g. a ' Slinky ') ; they may involve either longitudinal or transverse displacements of the atoms in a solid and their speed of propagation is given by $\sqrt{}$(appropriate elastic constant/density). An elastic stress pulse is produced in a solid whenever an externally applied force is changed suddenly or by the sudden change in local stress resulting from the initiation of a crack. Since

181

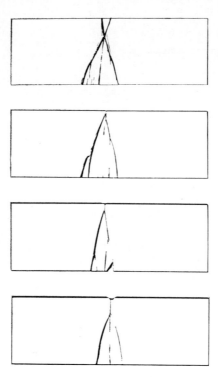

Fig. 6.19. Fracture branching in microscope slides. For three of the slides, fracture started at the edge and for the other (at top of photograph) fracture started at a point on the face of the slide about one-third of the width from the top edge.

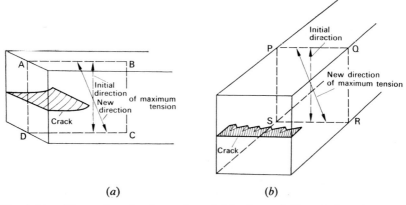

(a)   (b)

Fig. 6.20. Illustrating the formation of (a) ribs and (b) striations (see text).

182

even the terminal speed of a crack is less than the speed of elastic stress waves, a stress pulse can travel across a fracturing solid and be reflected back from the surface long before rupture is complete.

Rib-marks are probably due to the simultaneous interaction of the whole of a slowly propagating crack front with a stress pulse ; some pronounced ribs may mark positions where a stress pulse actually halted, momentarily, the propagation of the crack.

Wallner lines are usually rather smaller undulations in the crack plane and they often occur in groups of intersecting lines, fig. 6.17 c ; they are most easily seen in reflected light with a low power microscope which is slightly off-focus. The most common Wallner lines are produced by the interaction of the crack tip with stress pulses which are generated by the crack front itself as it passes an irregularity on the surface of the specimen. Pronounced Wallner lines can be produced by lightly grinding with emery cloth the surface of a glass rod and breaking it in bending ; the lines can be seen with a small hard lens, although a low power microscope with reflected light shows much more detail.

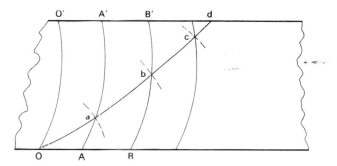

Fig. 6.21. Illustrating the formation of a Wallner line (see text).

Figure 6.21 illustrates the interaction between the crack and the stress pulse, which leads to the formation of a Wallner line. The plane of the drawing corresponds to the fracture surface of a plate of glass. The thin lines OO′, AA′, etc., represent positions of the crack front at successive, equal intervals of time ; the dotted lines, concentric circular arcs centred at O, mark the successive positions of a stress pulse originating at O, at the same times. Since the stress pulse travels faster than the crack front, the distances Oa, ab, etc., are greater, respectively, than OA, AB, etc. The superposition of the stress within the pulse on the stress generated by the externally applied fracture load may change the orientation of the maximum tension so that it is no longer perpendicular to the original crack plane, and then the crack tip will be deflected whenever it runs into material ' occupied ' by the pulse.

183

This will occur where the position of the crack front and the stress pulse coincide, i.e. at points like a, b and c. The locus of these points corresponds to a Wallner line, indicated in the figure by the heavy continuous line.

Measurement of crack speed using Wallner lines

The speed of a particular crack at different stages during its growth can be found, knowing the speed of stress waves, from a ' post-mortem ' examination of the fracture surface. The principle is very simple. In fig. 6.22 a Oaq represents a Wallner line in the fracture surface,

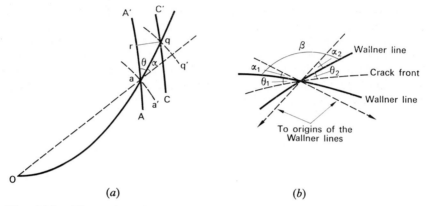

(a)   (b)

Fig. 6.22.   Illustrating the calculation of crack speed from orientation of Wallner lines (see text).

produced by a stress pulse generated at O ; aa′ and qq′ are the positions of the stress pulse at time t and $t + \delta t$; AA′ and CC′, respectively, are the positions of a section of the crack front at these same times.   The speed of the crack front is therefore $qr/\delta t = qa \sin \theta/\delta t$, but the speed of the stress pulse, c_1, is $q′a/\delta t$, hence the speed of the crack front is just

$$c_1 \sin \theta/\cos \alpha.$$

Since $\theta \rightarrow 0$ for slow cracks, the method is only applicable to cracks travelling at $\gtrsim \frac{1}{4}$ of the terminal speed, and for these faster cracks the orientation of the crack front, i.e. the line AA′, can seldom be established accurately from the fracture surface.   However, where two Wallner lines formed by pulses from separate sources intersect, both the speed and the orientation of the crack front can be found from the Wallner lines.   At the point of intersection there are two equations for the speed

$$\text{crack speed} = \frac{c_1 \sin \theta_1}{\cos \alpha_1} = \frac{c_1 \sin \theta_2}{\cos \alpha_2}$$

184

and $\theta_1 = (180 - \beta - \theta_2)$, where β is the angle between the two Wallner line segments at the intersection (fig. 6.22 b). These equations can be solved for the speed or the unknown angles θ_1 or θ_2 as required. Intersecting families of Wallner lines enable the speed to be determined at various positions on the fracture surface.

A much more elegant method for measuring the speed of cracks, first developed by F. Kerkhof, is to produce ripples in the fracture surface by continuously passing elastic stress (ultrasonic) waves into a specimen whilst it is loaded to fracture. By selecting the direction of propagation parallel to the applied tension and using transverse waves, undulations in the fracture surface can be produced which mark successive positions of the crack front at known time intervals, the period of the waves. Very high frequencies $\sim 10^6$ Hz and high ultrasonic powers are required to resolve the ripples produced by a crack travelling at the terminal speed.

These methods of determining the crack speed from fracture surface markings can only be used for materials which have very smooth fracture surfaces and even then they are restricted, essentially, to the smooth fracture zones. Some results for glass showing the acceleration up to a terminal speed, which occurs well before significant roughening sets in, are shown in fig. 6.23.

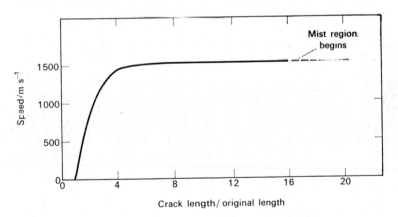

Fig. 6.23. Increase in crack speed in glass up to a limiting value.

Mirror, mist and hackle

As a crack expands under high stress, the energy release rate increases and rougher fracture surfaces are produced. The fracture surface of a glass rod broken in uniaxial tension is illustrated in fig. 6.24. There are three distinct zones on fracture surfaces of this kind : a smooth area, perpendicular to the applied tensile stress, extending from the

185

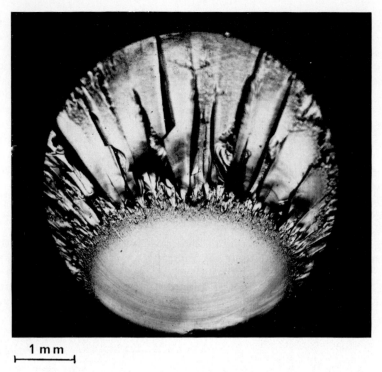

1 mm

Fig. 6.24.   Tensile fracture surface of a glass rod.

origin of fracture, an arc of rougher, stippled surface and a very heavily striated region in which the striations radiate as if from the fracture origin.   These fracture zones are designated, respectively, *mirror*, *mist* and *hackle*, fig. 6.25.   The mirror zone is usually marked by Wallner lines and other ripples.   The boundary between the mist and the hackle zones corresponds to the position at which branching occurs and, in fig. 6.24, the fracture surface beyond this boundary slopes away from the plane of the mirror zone, to form the surface of the crescent-shaped wedge of material which was isolated from the main fragments.

Progressive roughening of this kind is not confined to fracture surfaces formed under simple uniaxial tension nor is it found only on glass. An analogous sequence of zones can be seen on the fracture surfaces of all very brittle materials, although the detailed structure of each zone varies with the material ; for example, the ' mirror ' zone is not optically smooth for inhomogeneous materials like pottery or chocolate.   Except in the testing laboratory, fracture seldom occurs under a simple uniform tensile stress ; nevertheless crack propagation begins on a plane perpendicular to the maximum tension and mist and hackle zones are

186

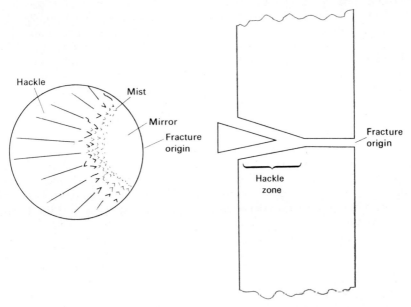

Fig. 6.25.   Forking and the fracture zones on a tensile fracture surface of a rod.

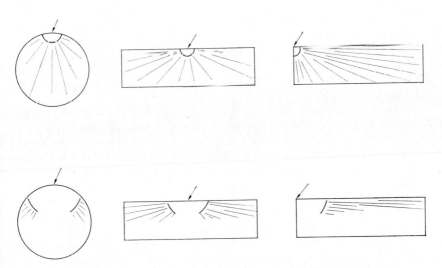

Fig. 6.26.   Schematic illustration of typical hackle distributions.   The arrows indicate the origin of fracture.   The top row shows the results of tensile fractures and the bottom row the results of bending fractures.

187

formed near the origin of fracture if this stress is high, i.e. provided the initiating flaw is not too severe. Once a long crack has been formed, it may continue to propagate into regions where the instantaneous stress is much lower, and then smooth fracture surfaces are formed. The shape of fracture zones changes with the distribution of stress ; for example a specimen with a moderately severe surface flaw, broken in bending, may have only small ' wings ' of mist and hackle, fig. 6.26, so that branching of the initial crack does not extend over the whole fracture front and no separate wedge of material is isolated from the main fragments.

The size of the mirror zone varies with the stress at which fracture occurred. It has been established empirically for several brittle materials of widely different structure (including glasses, ionic crystals, and inhomogeneous ceramic materials), that there is a very simple quantitative relationship between the ' mirror depth ', c and the tensile fracture stress, σ_b, viz.

$$\sigma_b^2 c = \text{constant.}$$

The value of the constant varies from one material to another ; for a soda–lime–silica glass it is $\sim 4 \times 10^{11} \text{ N}^2 \text{ m}^{-3}$.

No complete and entirely convincing explanation of this very simple relationship is available. Indeed, it is not yet clear even in qualitative physical terms why fracture branching occurs or even how it occurs. In principle, one or more of three different processes might be invoked to provide a mechanism which would lead to macroscopic forking : lateral division of the crack front as occurs in the formation of striations, advanced nucleation of secondary cracks slightly above or below the plane of the initial crack, or bifurcation at the tip of the initial crack. But it has still to be established which of these operates in practice for any given material.

Although the fundamental understanding of the cause of fracture zones is at a fairly primitive stage, the practical conditions required to form them are well established. Examination of the mist and hackle distribution on a fracture surface can yield, therefore, useful information about the cause and origin of fracture. The position of the mirror area reveals the origin of fracture ; the shape of the mirror zone may give a clue to the distribution of stress which existed near the fracture origin when failure occurred ; and the size of the mirror zone reveals the magnitude of the stress which caused the fracture. The distribution of mist and hackle has been used to establish that, for normal samples of glass, the serious flaws are invariably in the surface and also un-ambiguously to identify the particular flaws, such as the microscopic surface inclusion illustrated in fig. 6.9, responsible for the low strength of a given specimen. Unfortunately, the method cannot be used to trace the origin of fracture for ' flaw-free ' specimens of glass, because

188

(a)

Fig. 6.27. (a) Prince Rupert drops. A small gob of hot glass, produced by heating the end of a rod in an oxy-gas flame, is allowed to drop into water.

189

when failure occurs at these very high stress levels the specimen shatters : most of the thin very highly strained section of the sample is reduced to a fine powder.

6.8. *Toughened glass*

The flaws responsible for the ordinary low strength of glass are located

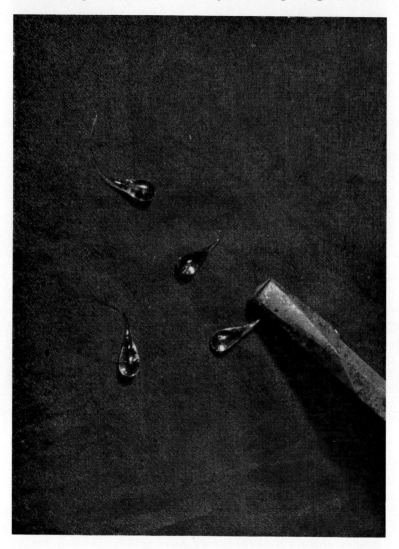

Fig. 6.27.   (*b*) The pear-shaped drop is very strong but the tail of the drop can be broken with the aid of pliers.

190

in the surface and they reduce the tensile stress which the material will support. These two characteristics allow articles of glass which are stronger than normal to be produced by creating 'permanent' compressive stresses in the surface layers where the danger lies. The

Fig.. 627.   (c) When the tail is broken, the whole drop disintegrates to a fine powder.

tensile stress required to fracture such an article is larger because it must first overcome the compressive stress before there is any tension on the surface flaws.

Glass which has been treated so that the surface layers are in compression is usually called *toughened glass*, although in this context ' toughened ' does *not* have the connotation of increased fracture energy. Indeed, for some types of toughened glass, once a crack has penetrated the compressive layers near the surface, it will propagate spontaneously even in the absence of an applied stress and repeated forking occurs so that the whole article shatters into small particles. It is for this characteristic, rather than for its greater strength, that toughened glass is selected for some applications, such as car windscreens, glass doors and large display windows.

Thermal toughening

The compressive stresses can be introduced thermally or chemically but the thermal method is by far the most widely used. The principle of this method has been known since the 17th century, in the form of ' Prince Rupert drops ' (fig. 6.27 pp. 189–91), but it has been exploited only during the past forty years or so.

In thermal toughening the glass object in its final form is heated in an electric furnace to a temperature close to its softening point and then it is removed from the furnace and the surfaces are cooled quickly, usually with a large number of jets of air. The surface layers rapidly become rigid, but the initial thermal contraction of the surface causes viscous flow in the central layers which are still close to the softening point of the glass. When the central layers eventually cool below the temperatures at which significant viscous relaxation can occur, a temperature gradient will exist through the thickness of the glass. This temperature gradient is removed when the object comes to thermal equilibrium at room temperature, and during this process, the hotter central layers must contract more than those near the surface. This differential contraction introduces internal stresses, which are compressive near the surface and tensile in the central layers. The variation of the stress across the thickness of a toughened plate of glass is roughly parabolic, fig. 6.28 ; thus the maximum compression in the surface is about twice the maximum tension in the centre, and the layer under zero stress occurs at a distance from the surface $\sim\frac{1}{5}$ of the thickness. There is of course, no net force on the plate, so that in fig. 6.28 the area A must equal the area $(B_1 + B_2)$.

Essentially, thermal toughening introduces balanced compressions and tensions within the thickness of the glass walls, by the removal at *low* temperatures of a temperature gradient from a stress-free sample ; the gradients can be introduced at *high* temperatures without causing a corresponding stress build-up, because the central layers are able to flow and so relieve the stress.   The magnitude of the stresses developed by this process depends on the thermal expansion coefficient of the glass and on the temperature gradient which can be established before viscous flow in the central layers becomes negligible.   This temperature gradient increases with the rate of removal of heat from the surfaces and with the thickness of the glass.   Higher stresses can be developed in a soda–lime–silica glass than in a low-expansion borosilicate glass : for a plate 6 mm thick and the normal cooling rate attainable with air jets, a compressive surface stress of ~ 100 MN m^{-2} can be produced in a soda–lime–silica glass with expansivity $\sim 9 \times 10^{-6}$ K^{-1} and a stress of ~ 40 MN m^{-2} in a borosilicate glass with expansivity $\sim 4 \times 10^{-6}$ K^{-1}.   Since the normal breaking strength is ~ 80 MN m^{-2}, these compressive stresses will raise the strength of plates of these glasses by about 2 and 1·5 respectively.

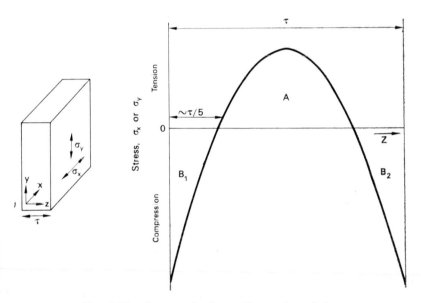

Fig. 6.28.   Stresses in thermally toughened glass.

Thermal toughening of flat or slightly curved plates $\gtrsim 6$ mm in thickness is widely employed.   For thinner plates the stress level, and therefore the increase in strength, produced when the surfaces are cooled by jets of air, is quite small but significant strengthening can be achieved if the rate of heat transfer from the surface is increased, for example by quenching the plate into a bath of liquid tin or Wood's metal.   Objects of more complex shape, especially those with varying wall thickness,

193

are more difficult to strengthen uniformly ; even for relatively simple hollow-ware (pans, bowls, beakers, etc.) it is difficult to arrange uniform cooling over both the inner and outer surfaces, although articles of this kind are often given some thermal toughening in order to improve their resistance to thermal shock.

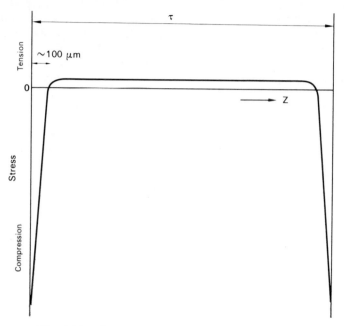

Fig. 6.29.   Stresses in chemically toughened glass.

Once a crack penetrates into the tension zone in thermally toughened glass it will propagate under the internal stress, even in the absence of externally applied forces.   If the internal tensile stress is large enough, the crack will accelerate and then branch.   Since the stress is *biaxial* (fig. 6.28), crack propagation can occur in any plane perpendicular to the surfaces of the plate.   Therefore both branches formed by the initial crack will propagate and will themselves branch ; this process continues until the whole plate is broken up into a large number of roughly rectangular fragments.   Many of the cracks propagating initially in the central tension zone do not extend to the surfaces of the plate and an externally applied force is required to separate the fragments.   This is why it may be necessary to punch a hole through a windscreen which has shattered spontaneously due perhaps to delayed fracture from an abrasion crack which penetrated just beyond the layer

194

of zero stress. The size of the fragments is determined by the magnitude of the internal stress : smaller fragments are produced as this stress is increased, essentially because the distance each crack travels before branches occur (i.e. the mirror ' depth ') decreases with increasing stress. The small, light fragments produced when toughened glass fractures are less dangerous than the large, sharp, pointed splinters often formed by ordinary plate glass. Toughened glass is therefore widely used as a safety glass ; the average fragment size can be controlled by varying the rate of cooling of the surfaces and the size required varies from one application to another.

Chemical toughening

Adjustment of the chemical composition of the surface layers of a glass object may also be used to introduce compressive stress. If the object is coated at a high temperature with a different glass which has a lower expansivity than the base glass, then, on cooling, the coating will be in compression. Alternatively, a similar effect can be brought about by modifying the surface composition of the original glass to produce a composition which has a lower expansivity. Recent interest has concentrated on a rather different method of chemical toughening which does not involve very high temperatures, so that there is no danger of distorting the shape of the original glass object. This method is to replace, at temperatures well below the annealing range, some of the original alkali metal ions in the surface layers of the glass by larger ones ; for example, some sodium ions are replaced by potassium or some lithium ions by sodium. This exchange of ions would lead to an expansion of the silica network but since the exchange is confined to the surface layers, the expansion is resisted by the inner layers of unmodified glass. This technique is known as *crowding* or *ion-stuffing*. High compressive stresses can be developed in the surface of a soda–lime–silica glass simply by soaking it in molten KNO_3 at $\sim 350°C$. The exchange of K^+ for Na^+ proceeds despite the increase in elastic strain energy which occurs as a result. In this particular case, however, there is no significant increase in strength, because the compression is confined to a very thin surface layer and the normal surface flaws penetrate beyond the compression zone. Thicker compression zones can be obtained in a reasonable time for selected glass compositions in which the diffusion rate of the alkali metal ions is appreciably higher ; several patents have been issued for compositions based on the mixed-alkali alumino-silica system of glasses which are suitable for strengthening by the ion-stuffing techniques.

With a suitable choice of base glass, ion exchange in the surface layers will produce a material with a tensile breaking strength ~ 1000 MN m^{-2}. Chemical toughening can, therefore, result in a very strong material, although it has not yet been widely employed.

There are two very serious limitations to the application of chemically toughened glass ; both stem from the fact that the compression zone is very thin and, consequently, the balancing tensile stress is very small and the tensile zone comes very close to the free surfaces, fig. 6.29. Toughened glass offers only a small improvement in resistance to concentrated loads and thus to abrasion damage ; the load on a small hard ball required to cause a surface crack is only slightly greater for toughened glass. The local tensile stresses required to nucleate a crack in the surface of ordinary glass near the area of contact with the ball are ~ 4000 MN m^{-2} so that an initial superposed compressive stress, due to toughening, even if it is as high as 1000 MN m^{-2}, will only increase the critical load by ~ 25 per cent. Because the compression zone in ion-stuffed glass is very thin, there is a serious risk that such cracks will penetrate into the tensile region. Although this need not result in immediate rupture, since the tension is much smaller than in thermally toughened glass, all the advantage of the compression layers is lost and the strength falls to that of ordinary glass with similar surface flaws.

The other serious limitation is that the low internal tension will not ensure fragmentation when fracture does occur : large sharp splinters may be produced. On the other hand, chemically strengthened glass with its low internal tension can be worked by sawing or drilling without risk of fracture, and there are few restrictions on the shape of an object which can be strengthened by this method.

CHAPTER 7
devitrification of glass and glass-ceramics

7.1. *Introduction*

ALL glasses are metastable ; their permanence at normal temperatures is due to the relative immobility of the ions in the structure, which precludes simultaneous movement of groups large enough to form the lower energy configurations of a crystalline structure. The ionic mobility is greater at high temperature and for silica based glasses, at ~ 800–$1000°C$, crystals will grow in the course of time ; the glass is said to *devitrify* (cf. *vitreous*, meaning glassy).

The production of useful articles of glass depends on the ability to carry the glass through the preparation and forming-stages without devitrification ; the growth of a few crystals would ruin the glass. Crystals in a glass can destroy the uniformity of viscous flow in the working temperature range and therefore spoil its blowing or moulding qualities : they may cause loss of transparency and may also produce catastrophically high stresses as the glass cools, because the composition and thermal expansion coefficients of the crystals and the residual glassy phase may be very different. The critical temperature range in which devitrification will occur, and also the rate at which it occurs within this critical range, varies from one glass composition to another. These are among the most important characteristics considered in the choice of composition for practical glasses.

It was discovered recently that a particularly useful new type of material can be produced by completely devitrifying suitable glasses. These new materials are known as glass-ceramics ; they have a structure similar to a traditional ceramic, i.e. randomly oriented crystalline particles bonded together by a glassy matrix, but in a glass-ceramic all the crystals are extremely small, $\sim 10^{-6}$ m across. The glasses used to produce glass-ceramics are chosen to have a high crystal growth rate over a convenient temperature range and special techniques are used to ensure that, when the glass is devitrified, crystal growth will start from a large number of separate points to produce a polycrystalline aggregate with very small crystal size. As yet these materials are rather expensive and most applications are somewhat exotic, e.g. rocket nose cones and radomes for high-speed aircraft ; some domestic ware has already appeared and is marketed in this country by J. A. Jobling Ltd., as Pyrosil. Major developments in glass-ceramics are to be expected and we are likely to see a variety of new forms in the near future.

o   197

7.2. *Crystal growth from the melt*

The growth of crystals in the mixture of six or even more different oxides present in a commercial glass-melt is a very complicated process, which has not been studied in great detail. However, the basic physical principles established for the crystallization of liquids consisting of a single component or a simple mixture of two components, serve as a useful general background.

Whenever a substance undergoes a change of phase, whether it is from vapour to liquid, from liquid to crystal or from one crystal structure to another, the change takes place progressively. In the change from liquid to crystal, even for a liquid which consists of a single element, all the atoms in the liquid do not simultaneously change their relative positions and take up the ordered arrangement corresponding to the crystalline structure, when the temperature falls to the melting point. Although, once the temperature is below the melting point, the crystalline arrangement has the lower energy (strictly, lower free energy in thermodynamic terms), simultaneous rearrangement of all the atoms would require a very large and prohibitive activation energy. The transformation takes place progressively : small groups of atoms change their relative position to form very small crystals at various places in the mass of liquid and these small embryonic crystals serve as nuclei upon which further growth can occur. So at any one time only a few atoms need acquire the activation energy to change their position with respect to their immediate neighbours. Growth of the initial crystal embryos is not, however, automatic ; unless they are greater than a critical size, which depends on the extent to which the liquid has been cooled below its melting point, they will re-melt rather than grow. Thus the crystallization of a liquid, or the condensation of a vapour, involves two distinct stages : the formation of stable nuclei and the growth of these into large particles of the new phase.

It may seem somewhat paradoxical that the very small crystal embryos should be unstable, even when the temperature of the liquid is below the melting point. The instability arises because the atoms in a very small crystalline particle have a significantly higher average energy than those in a large one. In a small particle a greater fraction of the atoms is in the surface and surface atoms have a higher energy than internal atoms because they do not have correctly positioned neighbours. So that, in macroscopic terms, the contribution *per atom* from the surface energy is greater for a small particle than for a large one. Suppose the particle is spherical, radius r, the interfacial surface energy (surface tension) is f_s and that the change in free energy when unit volume of crystal is converted into liquid is f_r. The change in free energy due to the formation of the spherical crystal nucleus is then

$$\Delta F = -\tfrac{4}{3}\pi r^3 f_r + 4\pi r^2 f_s.$$

198

For any given temperature less than the melting point, f_r is positive i.e. the crystal has the lower energy and, *provided r is large*, ΔF will be negative and the energy of the whole two-phase system will decrease as the particle increases in size. However, if r is very small, ΔF may be positive : the energy of the system may be larger when a small crystal forms even though the temperature is below the melting point. The value of f_r changes with temperature ; at the melting point, atoms have the same free energy whether they are in the interior of the liquid or of the crystal so that $f_r = 0$, for $T = T_m$. The variation of ΔF with r for various temperatures $\leqslant T_m$ is illustrated in fig. 7.1. If $T < T_m$, there is a critical radius, r_c beyond which the energy decreases continuously as r increases. This corresponds to the critical size of the nucleus. If a spherical nucleus attains a radius r_c, it will grow spontaneously but, to form a nucleus of this size, the free energy of the system must be increased temporarily by $(\Delta F)_c$. Because f_r increases as $(T_m - T)$ increases, the critical radius and the activation energy required to form a nucleus of critical size both decrease as the temperature falls.

The probability that energy $(\Delta F)_c$ will be available during the random

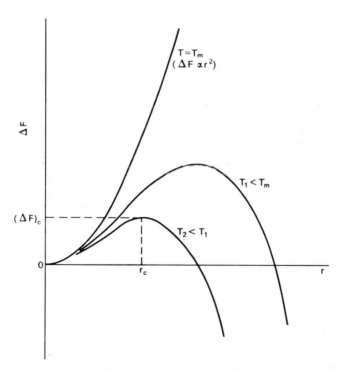

Fig. 7.1.   Illustrating the energy required for homogeneous nucleation.

thermal fluctuations is proportional to the Boltzmann factor, $\exp\left(-(\Delta F)_c/kT\right)$, so that the nucleation of stable crystalline particles becomes more probable as $(\Delta F)_c$ decreases. As the temperature is reduced still further below the melting point, the probability of nucleation will decrease again, because the effect of decreasing (kT) becomes greater than the decrease in $(\Delta F)_c$; this corresponds physically to the fall in the thermal energy associated with the atoms in the system. The probability, or rate, of homogeneous nucleation is zero at the melting point, rises to a maximum and then falls again to a very small value as the temperature is reduced.

The nucleation of stable crystalline particles may be aided by inhomogeneities in the liquid, such as small foreign particles which have neither melted nor dissolved in the liquid, or by rough edges on the surface of the containing vessel. These inhomogeneities may provide a surface upon which the atoms of the liquid can crystallize and which is initially greater than the critical size. The nature of the foreign particle or surface is very important; growth can occur much more readily if the atoms of the liquid are strongly attracted to those in the foreign surface and if the atomic spacing in the surface is very similar to the natural spacing in the crystal. This process is known as heterogeneous nucleation.

It is the difficulty of nucleating stable particles of the new phase which is responsible for supercooling of liquids, or the supersaturation of solutions and vapours. Homogeneous nucleation is usually much more difficult than heterogeneous nucleation and greater supercooling can be achieved if the initial phase is free from dirt particles. In an expansion cloud-chamber, the ions left in the track of a high-energy, charged, subatomic particle facilitate nucleation of liquid droplets in the supersaturated vapour. In practice it is usually necessary to operate the expansion diaphram a few times before looking for the tracks of charged particles, in order to clear the working space of dust particles which act as heterogeneous nuclei for the condensation of the vapour.

In principle the formation of a small stable crystal nucleus in a liquid below its melting point or in a saturation solution does not eliminate all problems of nucleation. A crystal consists of regular planes of atoms and large crystals grow from solution or from the melt so that their bounding surfaces are those planes in which the atoms are closely packed. As atoms are added to a crystal face, it will be necessary, periodically, to start a new layer, fig. 7.2. This is a two-dimensional nucleation and there will be a critical radius and an activation energy associated with the initiation of each new atomic layer, because atoms at the edge of the new two-dimensional island nucleus have a higher energy than those in the centre of the nucleus. The rate of two-dimensional nucleation should limit seriously the growth rate of crystals but in nearly all practical cases this does not happen; the linear rate of growth of crystals at low degrees of supercooling or of supersaturation is very much larger ($\gtrsim 10^{20}$!) than it should be. F. C. Frank first suggested a mechanism for the growth of crystals which would avoid the need to nucleate fresh layers. The presence of a particular type of lattice defect, a screw dislocation, effectively converts the separate atomic layers into a single spiral, rather like a spiral staircase. If a screw dislocation emerges at a surface of the crystal, it causes a small tapering step, fig. 7.3, and the addition of atoms to this step adds more turns to the spiral but does not eliminate the step. It has since been established by direct observation that such steps do occur on the growing faces of crystals of many different materials.

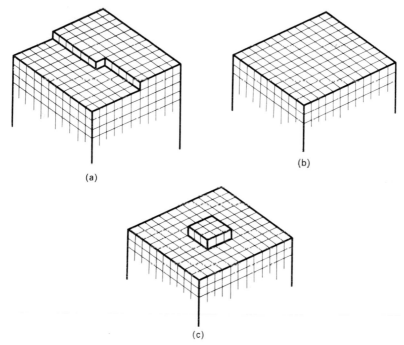

(a)

(b)

(c)

Fig. 7.2. Successive stages in the growth of a perfect crystal, using cubes to represent individual atoms. Once the partial layer in (a) is completed a closely packed surface (b) is produced ; further growth requires the nucleation of a new layer as in (c).

In practice, therefore, for simple liquids, homogeneous nucleation of the first very small crystals is the difficult stage in the crystallization ; once a stable nucleus has been formed, growth will occur rapidly. Even when there are no further nucleation barriers, the crystal growth rate will still vary with temperature. The linear rate of growth will be zero at the melting point, will increase rapidly as the supercooling increases but will eventually pass through a maximum and fall again when the temperature is so low that movement of the atoms or molecules in the liquid phase becomes very difficult. For liquids in which molecular movement is relatively easy, i.e. liquids of low viscosity, crystallization will normally be complete before these temperatures are reached, but for viscous liquids the nucleation rate and growth rate vary with temperature as shown in fig. 7.4.

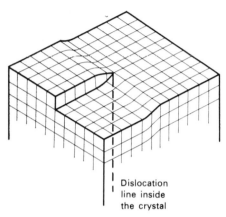

Dislocation
line inside
the crystal

Fig. 7.3.   Step on the surface of a crystal formed by the emergence of a screw
dislocation.   Growth can occur on this step which is self-perpetuating.

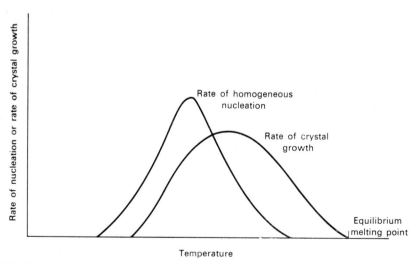

Fig. 7.4.   Expected variation with temperature of homogeneous nucleation
and crystal growth rate.

202

7.3. *Crystal growth in complex melts*

A liquid which is a homogeneous mutual solution of two or more different substances, e.g. water and alcohol, SiO_2 and Na_2O, Cd and Bi, does not have a single melting point above which the *equilibrium* form is liquid and below which it is crystalline. However, for any given pressure and composition of the mixture, there are two different temperatures which characterize the crystallization. The *liquidus* temperature is that above which no crystals exist in equilibrium and the *solidus* temperature is that below which no liquid exists in equilibrium. The difference between the liquidus and solidus temperatures is not trivial ; for a mixture of 90 per cent SiO_2 and 10 per cent Na_2O by weight the liquidus is $\sim 1520°C$ and the solidus $\sim 800°C$!

At any temperature between the solidus and liquidus temperatures, the equilibrium form is a mixture of liquid and one or more crystalline solids. The proportion of solid to liquid in this mixed-phase region changes as the temperature falls from the liquidus to the solidus, and the composition of the crystalline and liquid phases are, in general, not the same. The composition of the first crystals to appear as the liquid solution is cooled below the liquidus temperature depends on the composition of the liquid. When crystallization begins in the mixture 90 per cent SiO_2, 10 per cent Na_2O, pure silica separates out and the remaining liquid becomes progressively richer in Na_2O as the temperature is further reduced so that the proportion of crystalline material increases ; when the temperature reaches about $800°C$, a mixture of $(Na_2O . 2SiO_2)$ and SiO_2 crystals separate out and the equilibrium form is entirely crystalline but consists of a mixture of crystals with different compositions.

The details of the progressive crystallization in many different kinds of homogeneous liquid mixtures have been studied. The compositions and the structures of the crystalline phases which exist in equilibrium with one another and with a liquid phase are known as a function of the temperature and overall composition. Many mixtures of metals have been studied because their behaviour is of interest to metallurgists investigating the formation of alloys ; mixtures of minerals and oxides, because their behaviour interests geologists and because of the application to ceramics and glasses. All this detailed information refers to the equilibrium condition but, in order to achieve equilibrium when a liquid solution is cooled below its liquidus temperature, crystals of a composition different from the average composition of the liquid must be nucleated and grown. Both nucleation and growth will be much slower in a complex mixture than in a simple, pure liquid, because additional atomic or molecular rearrangements are usually required to change the local composition of a mixture. Crystal nucleation in a complex melt involves not merely simple rearrangement of the atoms but moving some atoms of one kind out of the way and at the same time arranging the others into the regular pattern of the crystal. It is often necessary for the composition of previously formed crystals to change in order to remain in equilibrium as the temperature falls. This change can occur very slowly by the diffusion of atoms within the crystal, but it may require a very long time at a particular temperature to reach the equilibrium configuration. Not surprisingly, therefore, the crystallization which occurs when complex melts are cooled at a normal rate seldom follows the whole equilibrium sequence. Thus most alloys and igneous rocks, although crystalline, are like glasses, in the sense that they are in a metastable state at normal temperatures ; the structure and composition of the crystalline phases present will change toward the equilibrium configuration if they are reheated to a temperature which is high enough to permit significant atomic migration.

7.4. *Devitrification of glass*

Normal glasses

In view of the importance of devitrification in the glass industry, there have been surprisingly few fundamental investigations of this process in glasses of commercial interest. Often the information available for a particular glass is limited to its liquidus temperature, the nature of the first crystals to appear when the glass is heated to selected temperatures below the liquidus and whether devitrification is ' easy ' or ' difficult '.

Although in principle a liquid could be supercooled sufficiently to form a glass if *either* the nucleation rate *or* the crystal growth rate were very small, in practice commercial glasses must have a very low crystal growth rate, since, on an industrial scale, elimination of heterogeneous nuclei is virtually impossible, particularly at the surface where the glass comes into contact with refractory materials and with ' dirt ' from the atmosphere.

Commercial glass compositions are almost invariably mixtures of oxides ; each of these mixtures has a very low liquidus temperature and a very high viscosity at this temperature. Necessarily, therefore, shaping operations are carried out at temperatures above the liquidus where there is no danger of devitrification and the formed article can be cooled through the critical temperature region relatively quickly. When particular shaping operations require glass at a high viscosity, e.g. in some automatic bottle-making machines or in a continuous sheet-drawing process, the composition of the glass used is specially adjusted to reduce the rate of crystal growth. This is one of the reasons for the variety in detailed composition of glasses produced for essentially the same end-product ; small adjustments or additions to the composition are made empirically to suit the particular forming processes used.

Experimental measurements of the crystal growth rate have been

204

made for a few commercial glasses and several simple binary glasses. In the most widely used technique, a sample is heated to a known temperature for a given time and then cooled rapidly so as to quench-in the condition which existed at the high temperature. Since crystal growth rates are very small and fall quickly to zero as the temperature is reduced, negligible changes occur during the rapid cooling and the size and nature of the crystals can be ascertained by microscopic and X-ray analysis at room temperature. These experiments have shown that the linear growth rate varies with temperature in the way expected, fig. 7.5. It is also found that the first crystals to grow, even in a simple glass, are not necessarily those which would be in equilibrium with a liquid phase at that temperature and composition. Neither are the crystals necessarily of simple composition or structure. The first crystals to appear when soda–lime–silica glasses devitrify have the composition ($Na_2O . 3CaO . 6SiO_2$); these crystals are not found elsewhere in nature and the compound has been named *devitrite*.

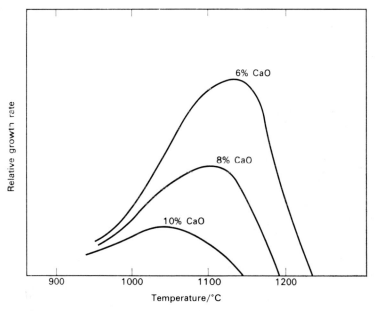

Fig. 7.5.   Observed crystal growth rates (length of crystals) in soda–lime–silica glasses, showing the effect of increasing CaO content.

Crystal growth always starts at the surface of normal glasses. It is believed that small dirt particles on the surface act as nucleation centres ; homogeneous nucleation in the bulk of the glass has never been observed. Total, or nearly total, crystallization of a normal glass requires a very long time and results in a coarsely polycrystalline material usually containing a mixture of crystals with different compositions.

Glass-ceramics

Although attempts to produce polycrystalline materials by devitrifying glass were made, by M. Réaumur, as long ago as 1739, methods for producing potentially useful materials have been developed only since about 1956. Réaumur successfully devitrified soda–lime–silica glass bottles, but the coarsely polycrystalline product was mechanically weak and the bottles often distorted during the heat-treatment. S. D. Stookey of the Corning Glass Works, U.S.A., discovered that a sample of photosensitive glass which had been exposed to ultra-violet light could be converted, without any marked softening or distortion, into a very fine-grain polycrystalline material by heat-treatment. The original discovery was due to accidental over-heating of a furnace in the course of experiments with photosensitive glasses ! Subsequent investigation has shown that the small colloidal particles of metal present in the exposed photosensitive glass can act as nuclei for the growth of crystals in the glass. The very large number and uniform distribution of these metallic particles ensured that crystallization started simultaneously from many different points and hence led to small individual crystals in the final aggregate.

Alternative methods of producing a fine dispersion of nuclei have since been discovered and a large number of patents have been issued covering glass compositions and nucleation catalysts suitable for producing glass-ceramics. The information which has been made available from the recent intensive research work in this field, serves to underline the complicated nature of the crystal growth process in glasses ; not only are the first crystals to appear of different composition from those which would be in equilibrium, but their composition and structure also depend on the method used to nucleate crystal growth and on the rate at which the glass is heated.

Special glasses are used to produce glass-ceramics and these are chosen for their relatively high rate of crystal growth and the desirability of the crystalline phases produced during devitrification. There is, of course, some conflict between the desire for a glass in which the crystal growth rate is high, so that devitrification to a glass-ceramic can be completed in an economically reasonable time, and the need to avoid devitrification during the working of the glass. Uncontrolled devitrification at this stage must be avoided or consistent, strong materials will not be obtained.

Glass-ceramics have been produced commercially from glasses based on $MgO–Al_2O_3–SiO_2$ and $Li_2O–Al_2O_3–SiO_2$ with titanium dioxide (TiO_2) added to the glass-melt to act as the nucleation catalyst during the devitrification process. Quite wide variations in the proportions of the major constituents are possible and other oxides are often included in the glasses to improve the melting or working characteristics of the initial glasses or to influence the crystal growth process.

After the glass has been shaped to the desired form, using conventional glass-shaping techniques, and annealed to remove any internal stresses, the finished article is reheated, usually in two stages. During the first stage, just below the softening point of the glass, the TiO_2 or other nucleating agent forms a finely dispersed colloidal precipitate, which will heterogeneously nucleate the growth of crystals from the glass when the temperature is raised. The first stage may take about an hour and then the temperature is raised slowly. Crystals grow as the temperature rises and this changes the composition of the remaining glass ; the rate of heating is controlled so that these changes will preserve the rigidity of the article when the temperature rises above the softening point of the original glass. The maximum temperature in the second stage must be below that at which crystals will re-melt, but high enough to achieve the required degree of crystallization in a short time, i.e. a few hours.

Initial glass type	Crystals which have been found in large numbers ; presence of particular crystals may depend on initial composition and devitrification treatment
$Li_2O–Al_2O_3–SiO_2$	$Li_2O . Al_2O_3 . 4SiO_2$ (β-spodumene) $\rbrace$ these may contain additional SiO_2 in solid solution $Li_2O . Al_2O_3 . 2SiO_2$ (β-eucryptite) $Li_2O . SiO_2$ (lithium metasilicate) $Li_2O . 2SiO_2$ (lithium disilicate) SiO_2 (as quartz or as cristobalite)
$MgO–Al_2O_3–SiO_2$	$2MgO . 2Al_2O_3 . 5SiO_2$ (cordierite) usually present for all proportions of components in initial glass SiO_2 (crystobalite) $MgO . SiO_2$ (clino estatite) $2MgO . SiO_2$ (forsterite)

Table 7.1. Crystals found in glass-ceramics.

Some of the crystalline compounds detected when $Li_2O–Al_2O_3–SiO_2$ and $MgO–Al_2O_3–SiO_2$ glasses were devitrified in laboratory experiments are listed in table 7.1. The electron micrograph in fig. 7.6 shows the very fine structure produced in a commercial $Li_2O–Al_2O_3–SiO_2$ glass-ceramic.

0·5 μm

Fig. 7.6.   Transmission electron micrograph of a fracture fragment of glass-ceramic (lithium-alumino-silicate) (C. R. W. Liley).

7.5. *Properties of glass-ceramics*

A ' glass-ceramic ' like a ' glass ' or a ' metal ' is a type of material and the properties of any one member of the class may differ widely from the properties of others.   A detailed account of the properties of the glass-ceramics which have been prepared and examined in the laboratory would fill another book ; here we shall just outline some of the main features in order to compare them with glasses.

A glass-ceramic is an aggregate of several different kinds of crystal in a glassy matrix, but the very small size and random orientation of the crystals make a glass-ceramic much more homogeneous than a traditional ceramic and result in macroscopic physical properties which are isotropic.   The irregularity of the inter-crystal boundaries makes a glass-ceramic opaque at optical frequencies although the true absorption may be quite small ; excellent transparency may be found at lower

208

frequencies (longer wavelengths). As might be expected for an aggregate of this kind, some physical properties may be determined primarily by one of the components present ; for example, the dielectric loss and the temperature at which the material distorts appear particularly sensitive to the amount and composition of the residual glass phase present, while the permittivity and thermal expansivity depend primarily on the nature of the predominant crystalline phases. By selecting the initial glass composition and devitrification treatment so as to vary the nature of the predominant crystals, materials can be produced with high thermal expansivity to match that of metals, or with almost no thermal expansion and hence with an extremely high resistance to thermal shock. As a group, glass-ceramics are mechanically three to four times stronger in service than glasses ; they are much harder and more abrasion resistant and are, therefore, less susceptible to surface damage. The $MgO-Al_2O_3-SiO_2$ glasses are particularly valuable starting points because they contain no alkali metal ions. These parent glasses have high electrical resistivity and low dielectric loss but the glass-ceramics produced from them have even higher resistivities and lower dissipation factors.

Although the technology of glass-ceramics has advanced rapidly in the past decade, the basic scientific understanding of their properties, and also of the processes involved in their production, is still very sketchy. The detailed mechanisms responsible for nucleating crystal growth, even with the commonly used TiO_2, are still controversial.

SYMBOLS

THIS list gives most of the symbols used in this text. In following the commonly selected notations, it happens that the same symbol is used with more than one meaning. However, there should be little chance of confusion in context. The list does not include all the subscripts which are used but each symbol is defined in the text as it is introduced into the argument.

C capacitance

c $\begin{cases} \text{specific heat capacity} \\ \text{speed of light} \\ \text{length} \end{cases}$

d $\begin{cases} \text{length} \\ \text{interatomic spacing} \end{cases}$

E $\begin{cases} \text{energy} \\ \text{electric field strength} \end{cases}$

F force

G shear elastic modulus

h the Planck constant

I electric current

K bulk elastic modulus

k the Boltzmann constant

$\left.\begin{array}{l} L \\ l \end{array}\right\}$ length

M molecular weight

m mass

N_A the Avogadro constant

n $\begin{cases} \text{number} \\ \text{refractive index} \end{cases}$

P $\begin{cases} \text{dielectric polarization} \\ \text{load} \end{cases}$

q charge

R $\begin{cases} \text{electrical resistance} \\ \text{stress-optical coefficient} \\ \text{characteristic length} \end{cases}$

r length

T $\begin{cases} \text{temperature in kelvin} \\ \text{transmittance} \\ \text{surface tension} \end{cases}$

t time

$V \begin{cases} \text{voltage} \\ \text{constringence} \end{cases}$

v ; v_x, v_y, v_z velocity and its cartesian coordinates

x, y, z cartesian coordinates

Y Young's modulus

$\alpha \begin{cases} \text{polarizability} \\ \text{extension} \end{cases}$

β spring constant = force per unit displacement

β_T absorption coefficient

$\gamma \begin{cases} \text{activation energy} \\ \text{logarithmic decrement} \\ \text{fracture energy} \end{cases}$

δ phase angle

ϵ_r relative permittivity

ϵ_0 permittivity of free space

η coefficient of viscosity

θ angle

$\lambda \begin{cases} \text{wavelength} \\ \text{mean free path} \end{cases}$

μ_r relative permeability

μ_0 permeability of free space

$\nu \begin{cases} \text{frequency} \\ \text{Poisson's ratio} \end{cases}$

$\rho \begin{cases} \text{electrical resistivity} \\ \text{density} \end{cases}$

σ ; $\sigma_x, \sigma_y, \sigma_z$ stress and its cartesian coordinates

σ_f yield stress

τ relaxation time

ϕ activation energy

ω angular frequency

Physical constants and conversion factors

The Avogadro constant, N_A : $6 \cdot 03 \times 10^{23}$ mol^{-1}
The Boltzmann constant, k : $1 \cdot 38 \times 10^{-23}$ J K^{-1}
The Planck constant, h : $6 \cdot 62 \times 10^{-34}$ J s
Electronic charge, e : $1 \cdot 60 \times 10^{-19}$ C
Electron rest mass, m_e : $9 \cdot 11 \times 10^{-31}$ kg
Proton rest mass, m_p : $1 \cdot 67 \times 10^{-27}$ kg
Speed of light in a vacuum, c : $3 \cdot 00 \times 10^8$ m s^{-1}
Permeability of free space, μ_0 : $4\pi \times 10^{-7}$ H m^{-1}
Permittivity of free space, ϵ_0 : $8 \cdot 85 \times 10^{-12}$ F m^{-1}
1 ångström (Å) $= 10^{-8}$ cm $= 0 \cdot 1$ nm

CHEMICAL SYMBOLS

The following elements may be found in various glasses or are used in glass-making.

Symbol	Element	Atomic number	Relative atomic mass (based on carbon 12)
Al	Aluminium	13	26·98
Ag	Silver	47	107·87
As	Arsenic	33	74·92
Au	Gold	79	196·97
B	Boron	5	10·81
Ba	Barium	56	137·34
C	Carbon	6	12·01
Ca	Calcium	20	40·08
Cd	Cadmium	48	112·40
Ce	Cerium	58	140·12
Cl	Chlorine	17	35·45
Co	Cobalt	27	58·93
Cr	Chromium	24	52·00
Cs	Caesium	55	132·91
Cu	Copper	29	63·54
F	Fluorine	9	19·00
Fe	Iron	26	55·85
H	Hydrogen	1	1·01
K	Potassium	19	39·10
Li	Lithium	3	6·94
Mg	Magnesium	12	24·31
Mn	Manganese	25	54·94
Na	Sodium	11	22·99
O	Oxygen	8	16·00
P	Phosphorus	15	30·97
Pb	Lead	82	207·19
Pt	Platinum	78	195·09
S	Sulphur	16	32·06
Sb	Antimony	51	121·75
Se	Selenium	34	78·96
Si	Silicon	14	28·09
Sn	Tin	50	118·70
Ti	Titanium	22	47·90
U	Uranium	92	238·03
Zn	Zinc	30	65·37
Zr	Zirconium	40	91·22

INDEX

215

Transparency 88ff
  (*see also* Absorption, optical)
Transmittance (definition) 89

Vacancies 58–9, 62–3
Viscosity
  and dashpots 135
  of glasses 9, 19, 21, 204
  reference temperatures 18, 19

Wallner lines 179, 181–5
Working point 19, 20

X-ray diffraction 6–8, 24

Yield stress 145, 174–5
Young's Modulus 126
  of glasses 128, 130, 165